Math Kangaroo USA
Grades 1 and 2
Volume 1

Questions and Solutions

Odd Years

2005–2023

Editor in Chief
Agata Gazal
Chief Editorial Officer for Math Kangaroo USA
Billings, MT

Reviewers
Joanna Matthiesen
Chief Executive Officer for Math Kangaroo USA
Granger, IN

Izabela Szpiech
Chief Financial Officer for Math Kangaroo USA
Chicago, IL

Kasia Nalaskowska
Chief Information Officer for Math Kangaroo USA
Aurora, IL

Agata Gazal
Chief Editorial Officer for Math Kangaroo USA
Billings, MT

Contributors
Maria Omelanczuk
Former CEO and President of Math Kangaroo USA
Oswego, IL

Professor Andrzej Zarach
Math Content Reviewer
East Stroudsburg University, East Stroudsburg, PA

David Zarach
Math Content Reviewer
East Stroudsburg, PA

Cover and Graphics Credit
Magdalena Teodorowicz
Chief Design Officer for Math Kangaroo USA
Cordova, TN

We would like to give special thanks to countless other people who contributed to the questions and solutions of this book since 2005, chiefly to the Math Kangaroo question writers from all over the world who are part of the AKSF organization (www.aksf.org), Math Kangaroo solution writers, Math Kangaroo USA competition organizers, and Math Kangaroo Alumni. We would also like to thank the hundreds of educators who gave us feedback on the questions and solutions and finally the tens of thousands of students that take the challenge each year. Thank you all for your help in developing this book.

www.mathkangaroo.org

Printed by Classic Printing & Thermography, Wood Dale, IL

This book contains the math problems (questions) presented in the Math Kangaroo competition in the odd years 2005–2023, as well as the solutions. It can be used at home and at school.

For additional copies of this book, please contact the publisher:
Math Kangaroo USA
info@mathkangaroo.org

ISBN 979-8-3507-1533-0

Preface

Early elementary school is an optimal time to spark a child's interest in mathematics. Introducing them to math riddles, puzzles, and creative logic questions can accomplish this in fun and enjoyable ways. This book presents 240 entertaining problems and solutions presented to 1st and 2nd grade students during the Math Kangaroo Competition odd years spanning 2005-2023, a total of 10 tests. Each test consists of 24 questions divided into easy, medium, and difficult categories. All questions were selected at the annual Kangourou sans Frontières meeting where mathematicians from 100 countries work together to choose the most engaging and age-appropriate questions for the annual Math Kangaroo Competition.

This easy-to-use resource book includes questions, pictures, and interesting solutions that will challenge children to use math and logic as a tool for understanding the world around them. Problem solving is a skill that all children use, sometimes without even knowing it. This book will help students practice their math skills that often involve logical reasoning and reflecting on the solutions.

We hope this book will be cherished by students who love mathematics, parents who like to study math with their children at home, and educators passionate about teaching unconventional and challenging math. Students will benefit from this book and find it both insightful and entertaining.

Joanna Matthiesen

President and CEO of Math Kangaroo USA
June 2023

Contents

Part I

Questions

Questions from Year 2005

Problems 3 points each

1. In the enchanted garden of the Green King, there are apple trees that grow golden apples. Every day, 5 golden apples become ripe on each tree, and at the end of each day they fall from the trees. Today, the Green Gardener picked up 20 ripe apples that fell under the trees last night. How many apple trees that grow golden apples are there in the garden?

(A) 4 (B) 5 (C) 6 (D) 7 (E) 8

2. Alma, Maria, Anne, and Michael had 2 apples each. Each one ate one apple. How many apples do they now have altogether?

(A) 1 (B) 2 (C) 4 (D) 6 (E) 8

3. Only one digit from 1 to 9 is repeated three times in this drawing. The rest of the digits are repeated twice. Which digit is repeated three times?

(A) 9 (B) 8 (C) 3 (D) 4 (E) 7

4. How many **different** digits can you see in the picture to the right?

(A) 3 (B) 4 (C) 5 (D) 6 (E) 9

5. What number is hidden under the question mark in the picture below (on the last car)?

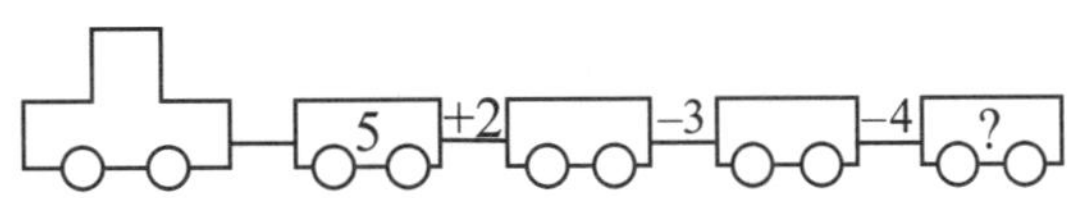

(A) 7 (B) 4 (C) 0 (D) 1 (E) 2

6. $5 - 4 + 3 - 2 + 1 = ?$

(A) 1 (B) 2 (C) 3 (D) 4 (E) 0

7. When Ann was born, Michael was 4. Now Ann is 3 years old. How old is Michael?

(A) 1 (B) 6 (C) 7 (D) 8 (E) 10

8. How many blocks were used to build the figure shown in the picture to the right?

(A) 7 (B) 12 (C) 13 (D) 14 (E) 16

Problems 4 points each

9. Which four beads below need to be added to the right side on the string of beads shown in the picture?

(A) ○●○○ (B) ●○○● (C) ○○○● (D) ○○○○ (E) ●●○○

10. How many more square tiles do we need to put on the kitchen floor to cover all of it? (See the picture.)

(A) 12 (B) 10 (C) 9 (D) 6 (E) 4

11. Helga is climbing stairs in such a way that she goes up 2 steps at a time. She is standing on the third step now. On which step will she be after she moves up 3 times?

(A) 9 (B) 1 (C) 6 (D) 5 (E) 8

12. Some pages are missing from an open book. On the left page you can see page number 12 and on the right page you can see page number 15. How many sheets are missing?

(A) 1 (B) 2 (C) 3 (D) 4 (E) 5

13. One hen lays one egg a day. In how many days will two hens lay 6 eggs?

(A) 1 (B) 3 (C) 6 (D) 9 (E) 10

14. There are two horses, one duck, one fish, an eagle, and a boy in a private garden. How many legs do they have altogether?

(A) 10 (B) 12 (C) 14 (D) 16 (E) 18

15. Anne has some apples. Maria has 2 apples more than Anne. Altogether they have 8 apples. How many apples does Anne have?

(A) 3 (B) 5 (C) 6 (D) 7 (E) 10

16. Which figure is next in this sequence: □ ⊞ ⊞ ... ?

(A)  (B) (C) (D) ⊞ (E) □

Problems 5 points each

17. Monika is 2 years old and Karl is 4. How old will Monika be when Karl is 11?

(A) 6 (B) 7 (C) 8 (D) 9 (E) 10

18. During the race, just before the finish line, I passed the runner who won third place. What place did I win?

(A) 1st (B) 2nd (C) 3rd (D) 4th (E) 5th

19. There are three weights, one weighing 1 kg, one weighing 4 kg, and one weighing 2 kg. They are shown on the balance scale in the picture on the left. There is only one weight of 1 kg on the balance scale in the picture on the right. What is the weight of the fruit in the basket?

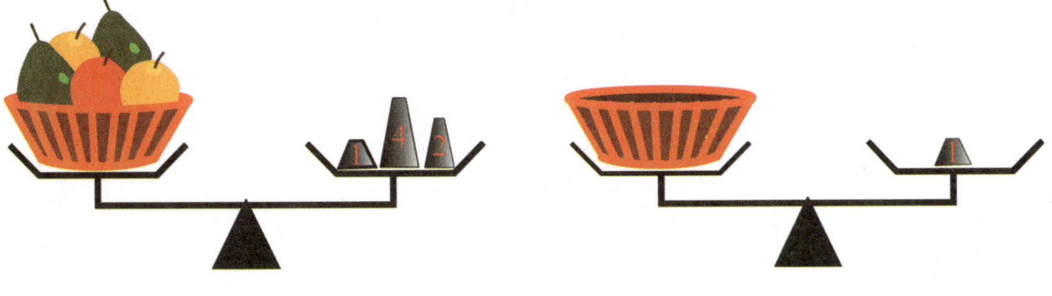

(A) 7 kg (B) 6 kg (C) 5 kg (D) 3 kg (E) 2 kg

20. Which set of signs + and − needs to be placed in the boxes to make the expression true?

$$4 \square 3 \square 2 \square 1 = 8$$

(A) +, −, + (B) −, +, − (C) +, +, − (D) +, +, + (E) −, +, +

21. The sum of two digits, one from inside the square and one from outside the square, is greater than 10. How many such pairs can we make?

(A) 19 (B) 11 (C) 6 (D) 24 (E) 18

22. What number is covered by "?" in the last picture below?

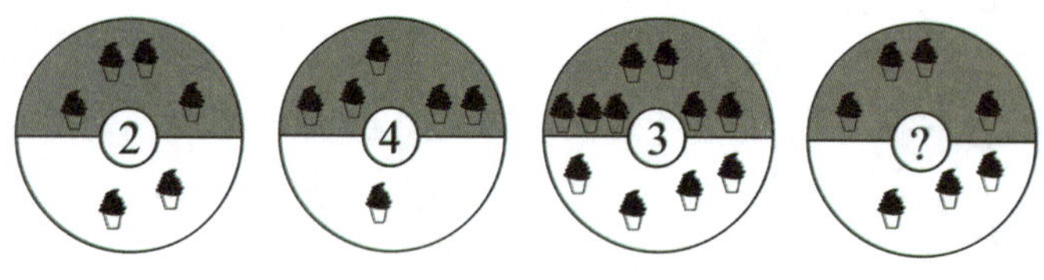

(A) 1 (B) 2 (C) 3 (D) 4 (E) 5

23. Hans fills the square table with numbers in such a way that the sum of the numbers in each column is 15, the sum of the numbers in each row is 15, and the sum on each diagonal is 15. What number will he put in place of "?" ?

(A) 1 (B) 3 (C) 2 (D) 6 (E) 9

24. A train has four cars in four colors: red, green, white, and yellow. The green car is not first and it is not last. The yellow car is not next to the white car and it is not next to the red car. The first car is white. What is the order of the cars in that train?

(A) white, green, red, yellow

(B) white, yellow, green, red

(C) green, yellow, red, white

(D) red, white, green, yellow

(E) white, red, green, yellow

Questions from Year 2007

Problems 3 points each

1. Which toy comes before the 5th toy?

(A)   (B)  (C)   (D)  (E) 

2. Which number is covered with the question mark?

(A) 1 (B) 2 (C) 3 (D) 4 (E) 5

3. How many bicycles can you see in the picture?

(A) 3 (B) 4 (C) 5 (D) 6 (E) 9

4. Find the value of each letter **R**, **A**, **C** and place the letters in increasing order according to their values. Which word do you get?

$7 - 2 = \mathbf{R}$ $1 + 3 = \mathbf{A}$ $10 - 8 = \mathbf{C}$

(A) **CAR** (B) **RAC** (C) **ARC** (D) **CRA** (E) **ARAC**

5. Which elf does not appear in the big picture?

(A)  (B)  (C)  (D)  (E) 

6. From the number of 101 Dalmatians from the animated film, take away the number of Snow White's seven Dwarves. What number do you get?

(A) 100 (B) 98 (C) 96 (D) 94 (E) 90

7. There are 5 kinds of dinosaurs in a bag. Of which kind is there the most?

(A)  (B)  (C)  (D)  (E) 

8. The road sign shows the distance between villages. How far is it from Behaiesti to Mormaiesti?

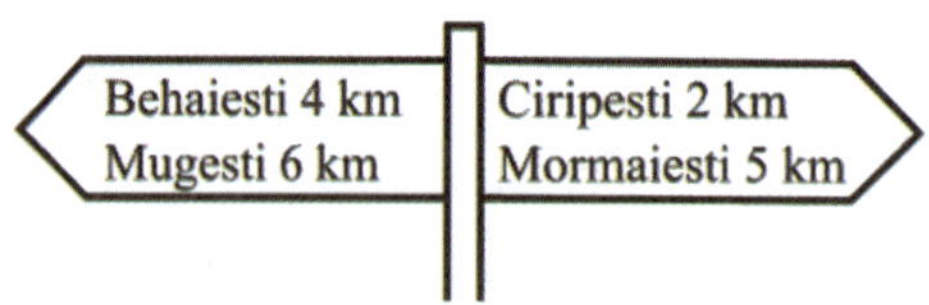

(A) 9 km (B) 6 km (C) 8 km (D) 11 km (E) 17 km

Problems 4 points each

9. Which letter is missing from each of the words below?

SCHOL BOK PRBLEM QUESTIN

(A) A (B) E (C) O (D) I (E) U

10. The picture can help you see how old each child is. Which sentence is true?

(A) Maria is 6 years old.

(B) Maria is 7 years old.

(C) Cristi is 9 years old.

(D) Cristi is 7 years old.

(E) Cristi is 8 years old.

11. When we left, we had filled the gas tank (50 liters). When returning, the indicator pointed as shown in the picture. How many liters did we use?

(A) 10 (B) 20 (C) 30 (D) 35 (E) 40

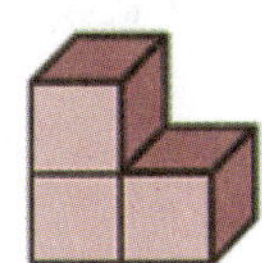

12. Michael glued 3 cubes together, as shown in the picture, and then he covered all of them with paint. How many squares did he paint?

(A) 3 (B) 7 (C) 10 (D) 14 (E) 16

13. Which of the children will reach the rabbit? Follow the arrows and help them!

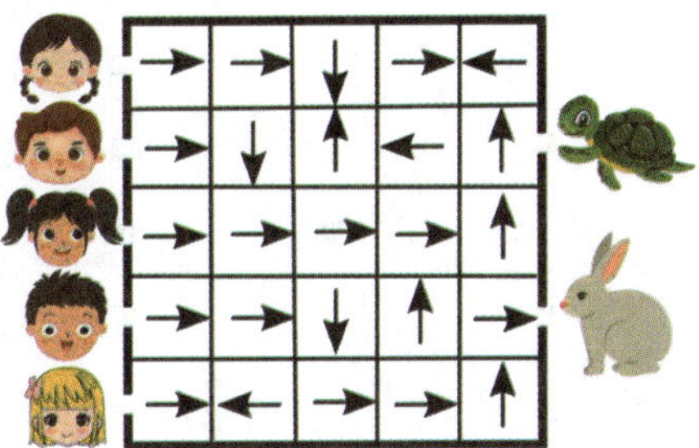

(A)  (B)  (C)  (D)  (E) 

14. What digit does the apple represent?

(A) 1 (B) 2 (C) 4 (D) 6 (E) 8

15. The clowns at a circus are standing one behind another. Vivi said to Bibi, who is standing behind him, "Behind me, 4 coworkers are standing." Bibi answered, "In front of me, only 3 coworkers are standing." How many clowns are standing in line?

(A) 7 (B) 8 (C) 9 (D) 10 (E) 12

16. If 2 dolls and 1 car cost \$8, and 1 doll and 2 cars cost \$7, how much do 3 dolls and 3 cars cost?

(A) \$15 (B) \$16 (C) \$21 (D) \$24 (E) \$14

Problems 5 points each

17. What is the sum: $3 + 7 + 2 + 8 + 1 = ?$

(A) 12 (B) 14 (C) 17 (D) 21 (E) 27

18. Dan left his home at 8:00 a.m. He left school at 12:30 p.m. The walk home takes 30 minutes. How long was he gone from his house?

(A) 4 hours (B) 4 hours 30 minutes (C) 5 hours
(D) 5 hours 30 minutes (E) 6 hours

19. In a certain number, the first digit is greater than the second digit by 2, and the second digit is greater than the third digit by 3. It is one of the numbers listed below. What number is it?

(A) 53 (B) 530 (C) 233 (D) 521 (E) 431

20. Simona is younger than Victor, but older than Tibi. Alice is younger than Tibi, but older than Barbu. Who is the oldest of them all?

(A) Barbu (B) Alice (C) Tibi (D) Simona (E) Victor

21. Which number should replace the question mark in the pyramid?

?
11 13
4 ... 6
1 3 4 2

(A) 10 (B) 14 (C) 22 (D) 24 (E) 34

22. What number is missing in the picture?

9 0 2
209 902 290 ?

(A) 210 (B) 202 (C) 92 (D) 308 (E) 920

23. Dan was given a bag of fruit-flavored candy (lemon—yellow, orange—orange, strawberry—red, mint—green, cocoa—brown). The bag contains 20 pieces of candy, the same number of each color. Dan first ate only all the yellow, red, and orange ones. How many pieces of candy are left in the bag?

(A) 5 (B) 6 (C) 15 (D) 12 (E) 8

24. Andrea needs an hour to get to the shopping center. If she leaves at 4 p.m., she gets there half an hour after the store closes. If she leaves at 8 a.m., she gets there half an hour before the store opens. What hours is the shopping center open?

(A) 7:30 to 4:30 (B) 8:30 to 5:30 (C) 7:30 to 5:30
(D) 8:30 to 4:30 (E) 9:30 to 4:30

Questions from Year 2009

Problems 3 points each

1. The figure shown in the picture on the right is built out of identical wooden cubes. How many cubes were used to build it?

(A) 12 (B) 8 (C) 9 (D) 10

2. What is the sum of the digits in the number 2009?

(A) 7 (B) 11 (C) 12 (D) 18

3. The picture shows the birthday cakes of Ala, Ola, Ela, and Ula.

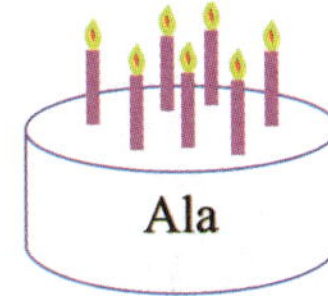

Which of the girls is the oldest?

(A) Ala (B) Ola (C) Ela (D) Ula

4. Which plate has fewer apples than pears?

(A)

(B)

(C)

(D)
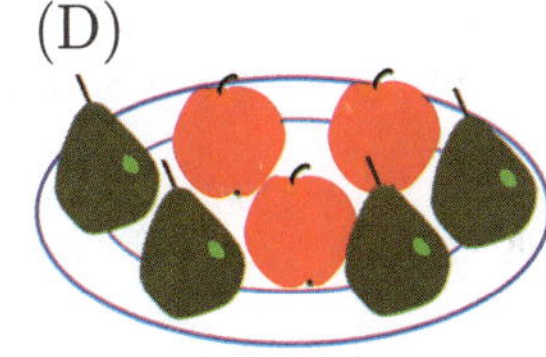

5. In the table shown in the picture, Ola wrote four numbers which have a sum of 50. What number is the butterfly sitting on?

5	
20	17

(A) 18 (B) 8 (C) 9 (D) 13

6. Peter has 12 toy cars, and Paul has 4 more toy cars than Peter. How many toy cars do Paul and Peter have together?

(A) 28 (B) 16 (C) 48 (D) 20

7. After Snow White, the Prince and the 7 Dwarfs ate one apple each, there were 4 apples left in the basket. How many apples were there in the basket before they ate any?

(A) 8 (B) 9 (C) 13 (D) 14

8. The duck in the picture is subtracting some numbers. Two butterflies landed on the picture, each covering up the same digit. What digit are the butterflies covering?

(A) 3 (B) 6 (C) 8 (D) 4

Problems 4 points each

9. On the last day of school, a father took his three children to the circus. How much did the tickets for all four of them cost?

TICKET BOOTH	
Child ticket	$9
Adult ticket	$12

(A) $48 (B) $21 (C) $39 (D) $30

10. Anna performed two operations. She put stickers over some of the numbers—the same stickers over the same numbers.

$$21 - 7 = \square$$
$$2 \times \square = \square + 1$$

What number can be found under the sticker?

(A) 15 (B) 14 (C) 25 (D) 27

11. The doctor gave a certain medicine to Ada. He gave her 60 pills, and told her to take one pill each day. Ada began taking the pills on a Monday. What day of the week will it be when she takes the last pill?

(A) Monday (B) Tuesday (C) Wednesday (D) Thursday

12. Sophie's mother bought 6 identical boxes of crayons. Sophie took out all the crayons from two of the boxes—there were 18 crayons. How many crayons did her mother buy altogether?

(A) 26 (B) 54 (C) 24 (D) 108

13. Tom is 2 inches taller than Peter, and 5 inches taller than Paul. How many inches taller is Peter than Paul?

(A) 7 inches (B) 3 inches (C) 10 inches
(D) Paul is taller than Peter.

14. Eva drew 6 flowers, and Ola drew 4 hearts. Ilona drew 3 times fewer flowers than Eva, and Ilona drew 2 hearts more than Ola. Which of the pictures below did Ilona draw?

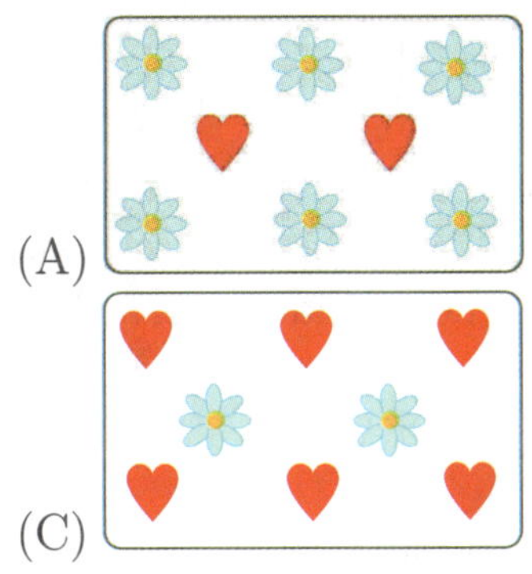

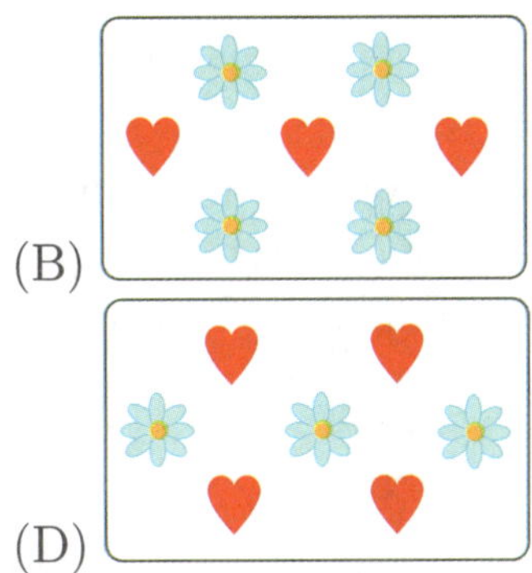

(A) (B) (C) (D)

15. What is the difference between the largest and the smallest of the numbers that fell out of Mr. Mouse's briefcase (see the picture)?

(A) 3 (B) 5 (C) 7 (D) 9

16. ET must have the clock which shows exactly 6:00. Which clock will he choose?

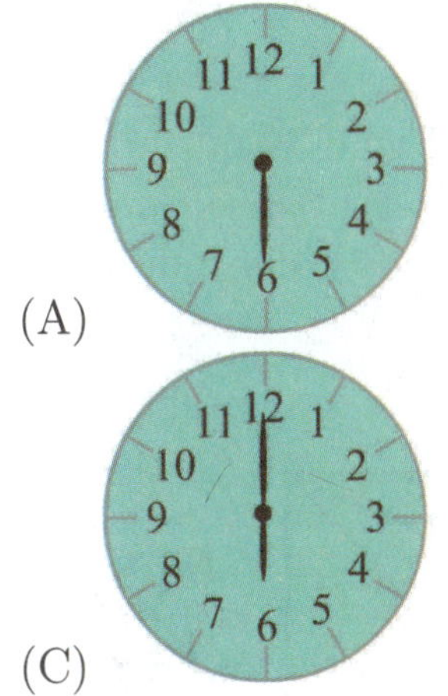

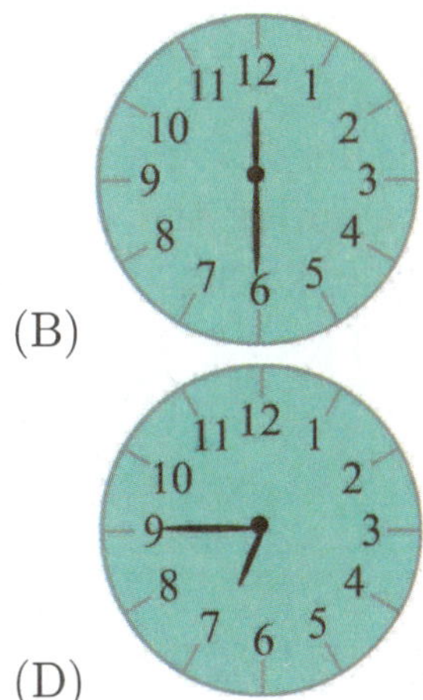

(A) (B) (C) (D)

Problems 5 points each

17. In a certain zoo, in the part that houses apes and monkeys, there were 19 animals. Among them were 4 chimpanzees and 3 baboons. The rest of the animals were gorillas, which were placed in three cages, the same number of gorillas in each cage. How many gorillas were in one of the cages?

(A) 5 (B) 4 (C) 3 (D) 6

18. Right now, John is 4 years old, and his father is 26 years old. How old will John's father be when John is 3 times as old as he is now?

(A) 78 (B) 38 (C) 42 (D) 34

19. Grandma made some cheese dumplings and some blueberry dumplings. Altogether, she made 31 dumplings. If she had made 11 more cheese dumplings, then there would be the same number of blueberry dumplings as cheese dumplings. How many cheese dumplings did grandma make?

(A) 10 (B) 21 (C) 20 (D) 15

20. Eva bought 2 identical notebooks. She had $4 left over. If she wanted to buy two more notebooks like these, she would need $2 more than she started with. How much did one notebook cost?

(A) $2 (B) $10 (C) $6 (D) $3

21. Adam, Matt, Paul and Tom were looking at their stamp collections. They found that Matt had more stamps than Paul, and Tom had fewer stamps than Adam. Also, Tom did not have the smallest number of stamps. Which of the boys had the smallest number of stamps?

(A) Adam (B) Matt (C) Paul
(D) This cannot be determined.

22. Father was gathering mushrooms for 2 hours. During the first hour, he found 39 mushrooms. How many mushrooms did he gather during the second hour, if it is known that in 40 minutes mother cleaned all the mushrooms father gathered, and she cleaned 7 mushrooms every 5 minutes?

(A) 39 (B) 17 (C) 74 (D) 56

23. Which triangle is Tringhi's twin brother (see the picture)?

(A) 3 (B) 4 (C) 5 (D) 6

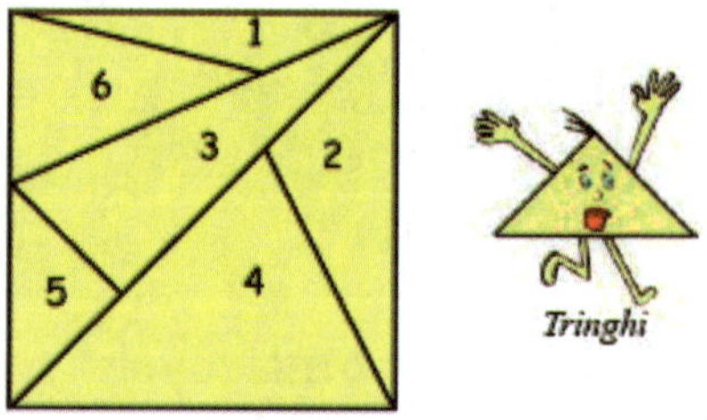

24. How many different three-digit numbers can we make that have three different digits using the digits 1, 2 and 3?

(A) 3 (B) 6 (C) 8 (D) 12

Questions from Year 2011

Problems 3 points each

1. Consecutive natural numbers were placed in the cells of the table below. What number is missing from the middle cell?

1	2	?	4	5

(A) 0 (B) 1 (C) 3 (D) 6

2. $6 + 2 =$

(A) 5 (B) 6 (C) 7 (D) 8

3. Sharon had 10 dolls. She gave Betty one of her dolls. How many dolls does Sharon have now?

(A) 6 (B) 7 (C) 8 (D) 9

4. There are 2 boys and 2 dogs and nobody else on the playground. How many legs are there on this playground?

(A) 12 (B) 10 (C) 8 (D) 4

5. Which month sometimes has only 29 days?

(A) January (B) February (C) March (D) April

6. 7 students and a teacher are ready for a snack. There are 7 glasses of milk, 8 candy bars, and 1 cup of coffee ready for them. Each student will have the same snack. How many candy bars will the teacher get with his coffee?

(A) 0 (B) 1 (C) 2 (D) 3

7. What is the sum of the digits in the number 2011?

(A) 202 (B) 31 (C) 4 (D) 13

8. Katie's doll is wearing a dress, has two braids, and is holding one flower in her hand. Which picture shows Katie's doll?

(A) 

(B) 

(C) 

(D)

Problems 4 points each

9. At the end of the skiing season, there were 12 pairs of ski boots left at the store. How many ski boots counted one by one were left at the store?

(A) 6 (B) 12 (C) 24 (D) 4

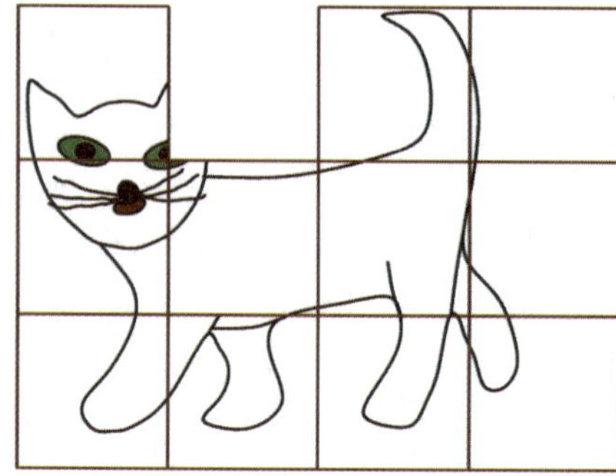

10. The picture to the right shows a puzzle with one piece missing. Which of the pieces below needs to be added to the puzzle in order for it to make a picture of a cat?

(A) (B) 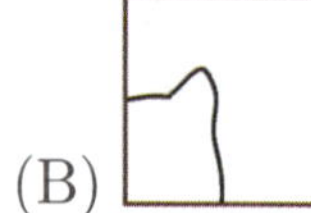 (C) 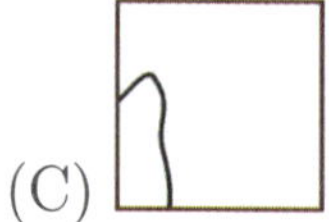 (D) 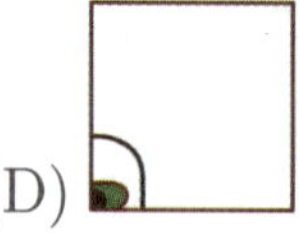

11. Today is 3/12/2011. The pictures below show food items and their expiration dates. Which of the items is past its expiration date?

(A)  9/15/2011

(B)  3/4/2012

(C)  7/11/2011

(D)  2/25/2011

12. In 36 years, Mark's grandmother will celebrate her 100th birthday. How old is Mark's grandmother now?

(A) 74 (B) 64 (C) 66 (D) 36

13. Anne has several dogs and 4 cats. The number of her cats' ears is equal to the number of her dogs' paws. How many dogs does Anne have?

(A) 8 (B) 2 (C) 4 (D) 6

14. To find her toy, Marie needs to follow the path which is marked by the following signs in this order: △, ♡, ◇, △, ♡, ◇, △, ♡, ◇. Which toy belongs to Marie?

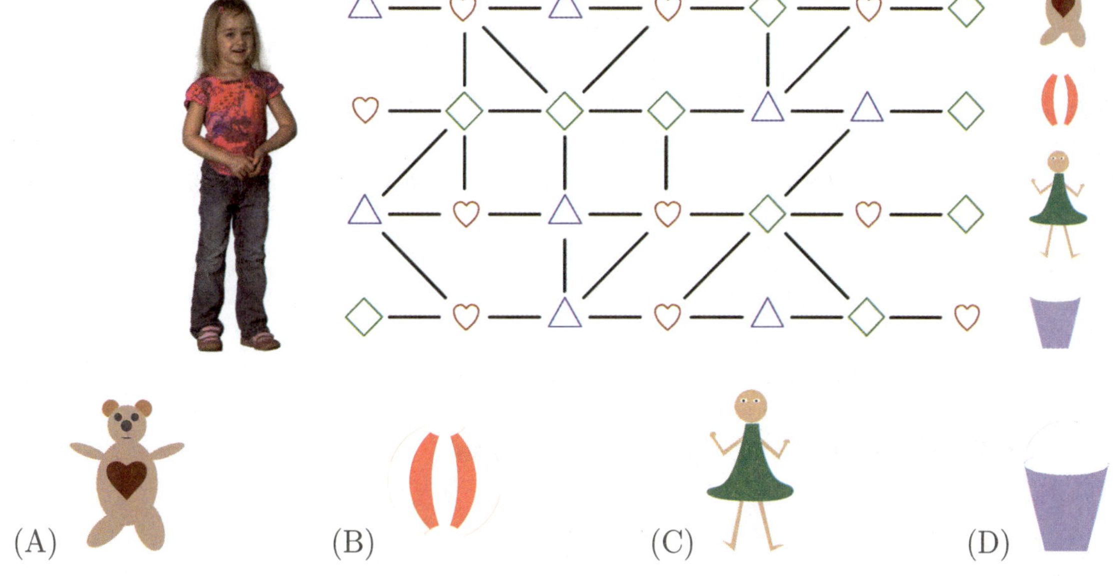

(A) (B) (C) (D)

15. The picture below shows part of a train schedule. Right now, it's 8:45. Mr. Smith will go from Chicago to Indianapolis on the next train. The trip will take 2 hours and 45 minutes. What time will Mr. Smith arrive in Indianapolis?

CHICAGO – Departures					
INDIANAPOLIS	6:55	8:30	9:15	11:15	12:50

(A) 11:30 (B) 12:00 (C) 11:15 (D) 12:15

16. Katie bought three identical pencils, two identical pens, and two identical erasers, and paid \$11.60. Hannah bought one pencil, two pens, and two erasers, and she paid \$8.40. How much does one pencil cost?

(A) \$1.20 (B) \$1.50 (C) \$1.60 (D) \$3.20

Problems 5 points each

17. Natalie folded a piece of paper in half and cut out a shape, as shown in the picture to the right. Which of the pictures below shows the piece of paper after it was unfolded?

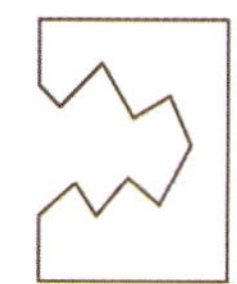

(A) 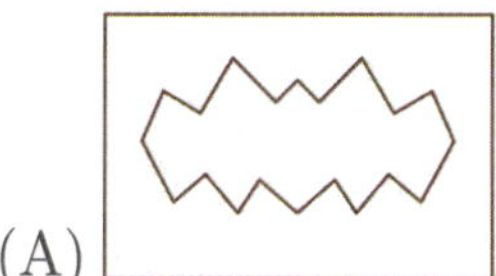 (B) 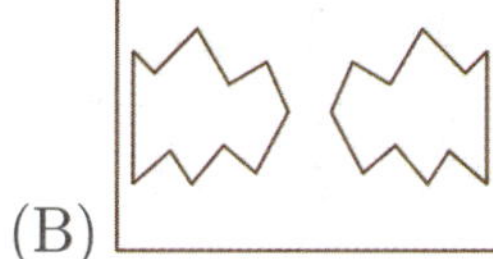 (C) 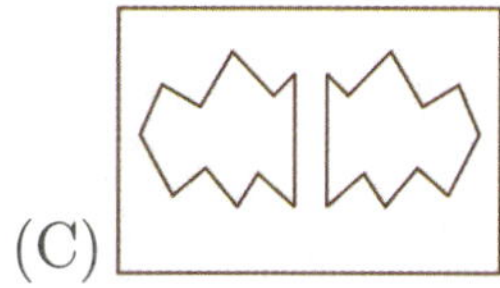 (D) 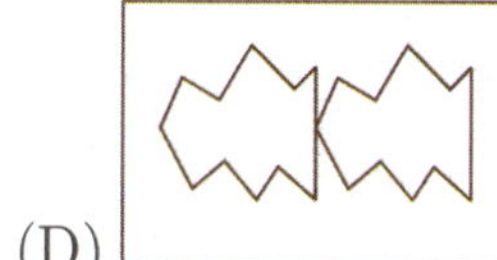

18. Mr. and Mrs. Taylor have three daughters. The youngest is 5 years old. The middle daughter is 4 years younger than the oldest daughter and 6 years older than the youngest daughter. How old is the Taylors' oldest daughter?

(A) 10 (B) 11 (C) 9 (D) 15

19. The flowers in the flowershop were kept in three vases. There were 16 flowers in the first vase, 11 flowers in the second vase, and 17 flowers in the third vase. The owner decided to sell only bouquets of 5 flowers each. After selling some bouquets, she noticed that she did not have enough flowers to make another bouquet. How many flowers did she have left?

(A) 1 (B) 2 (C) 3 (D) 4

20. Simon has two identical aquariums. There are 26 quarts of water in one and 42 quarts of water in the other. How many quarts of water does Simon need to pour from the second aquarium into the first in order to have the same amount of water in both?

(A) 6 (B) 16 (C) 10 (D) 8

21. Fido the Dog, Philemon the Cat and 4 monkeys together weigh 24 lbs. Fido and one monkey together weigh 11 lbs. Philemon and 2 monkeys together weigh 1 lbs less than Fido and one monkey weigh together. Each of the monkeys weighs the same. How much does Philemon weigh?

(A) 3 lbs (B) 4 lbs (C) 5 lbs (D) 6 lbs

22. Anita, Clara, Michael, and Daniel had an apple eating contest. The person who ate the most apples won. Daniel ate more apples than Clara, and Michael ate fewer apples than Anita. We also know that Daniel did not win. Who ate the most apples?

(A) Anita (B) Clara (C) Michael (D) We cannot know.

23. What number do we need to put in the first square in order to get 100 as the result after doing all the operations shown below?

□ —×4→ ○ —+2→ □ —×2→ ♡ 100

(A) 11 (B) 9 (C) 14 (D) 12

24. Paul and Jon were building using identical cube blocks. Paul made the building shown in Picture 1. Picture 2 shows Paul's building as seen from above. Picture 3 shows Jon's building as seen from above. (Note: The numbers in each square indicate how many blocks are placed one on top of another in that place.) Which of the answers shows Jon's building?

Picture 1.

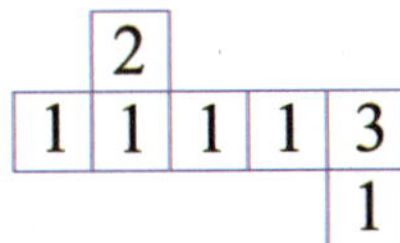

Picture 2.

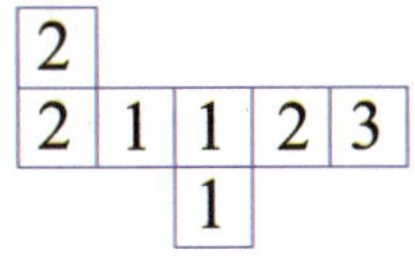

Picture 3.

(A) 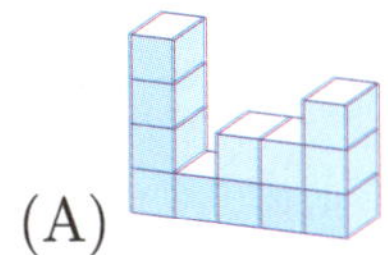 (B) 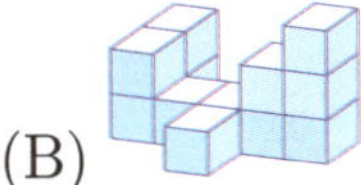 (C)  (D) 

Questions from Year 2013

Problems 3 points each

1. Which digits are missing on the right?

(A) 3 and 5 (B) 4 and 8 (C) 2 and 0
(D) 6 and 9 (E) 7 and 1

2. There are 12 books on a shelf and four children in a room. How many books will be left on the shelf if each child takes one book?

(A) 12 (B) 8 (C) 4 (D) 2 (E) 0

3. Which of the dresses has less than 7 dots, but more than 5 dots?

(A)  (B) 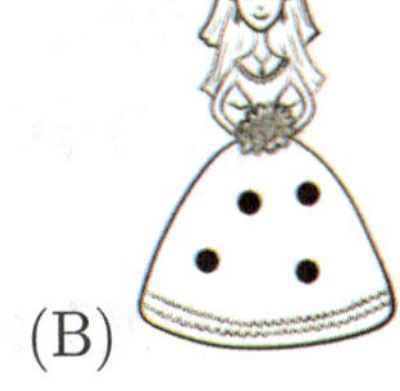 (C)  (D) 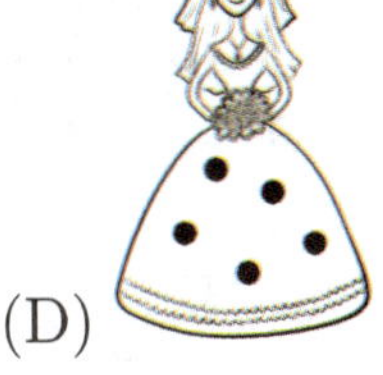 (E) 

4. There are white, gray, and black kangaroos. Which picture has more black kangaroos than white kangaroos?

(A) 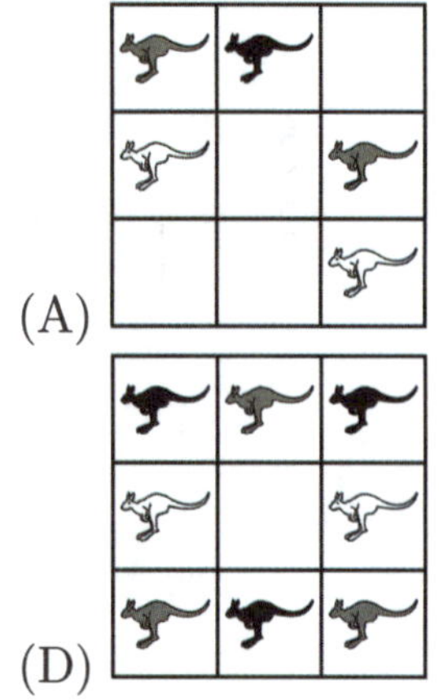 (B) 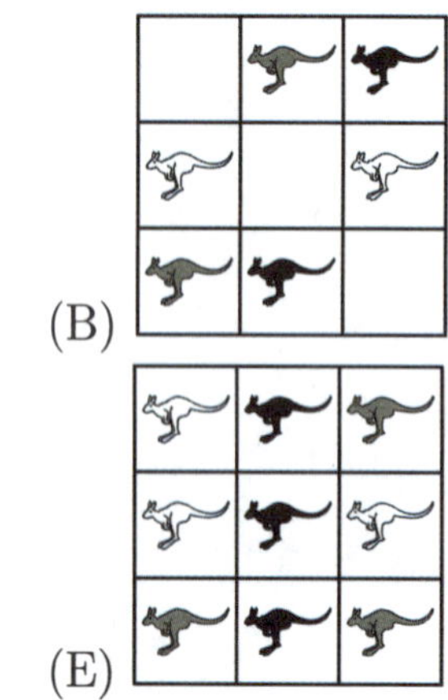 (C) 

(D) (E)

5. How many more bricks are there in the larger stack?

(A) 4 (B) 5 (C) 6 (D) 7 (E) 10

6. The picture shows a path made of square tiles. How many tiles fit in the area inside?

(A) 5 (B) 6 (C) 7 (D) 8 (E) 9

7. Lotta cuts a big piece out of a cake. Which one?

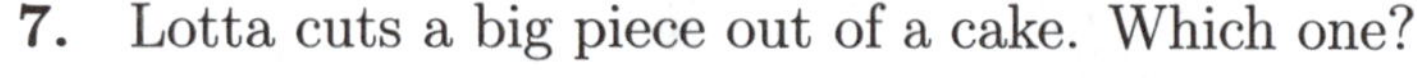

(A)  (B) 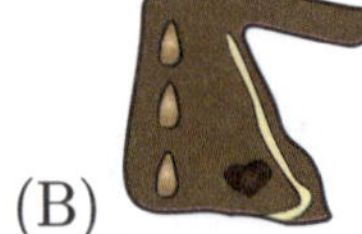 (C)   (D)  (E) 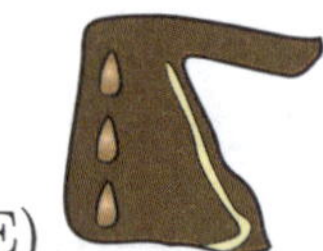

8. Ann has . Barb gave Eve . Jim has . Bob has . Who is Barb?

(A)  (B) 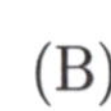  (C)   (D)  (E) 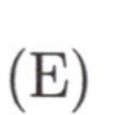

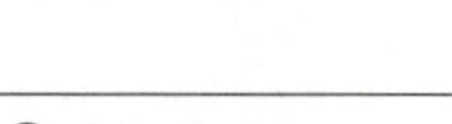

Problems 4 points each

9. Father gives 5 apples to each of his three children. Ana gives 3 apples to Sanja and then Sanja gives half of her apples to Mihael. How many apples does Mihael have now?

(A) 4 (B) 5 (C) 7 (D) 8 (E) 9

10. George has 2 cats of the same weight. What is the weight of one cat if George weighs 30 kilograms?

(A) 1 kilogram (B) 2 kilograms (C) 3 kilograms
(D) 4 kilograms (E) 5 kilograms

11. Which kind of square appears most often?

(A) (B) (C) (D) (E) all equal

12. How many carrots can the rabbit eat walking freely in this maze?

(A) 7 (B) 8 (C) 9 (D) 15 (E) 16

13. Cat and Mouse are moving to the right. When Mouse jumps 1 tile, Cat jumps 2 tiles at the same time.

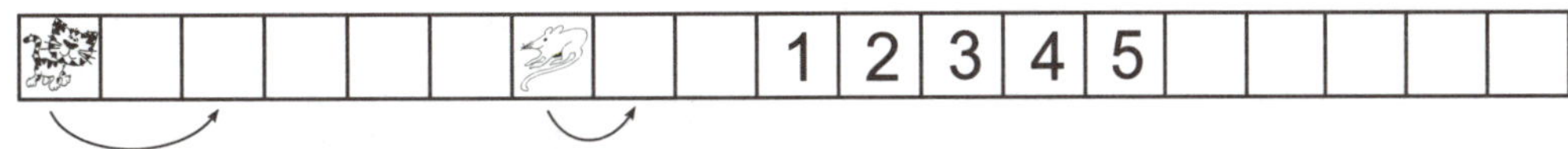

On which tile does Cat catch Mouse?

(A) 1 (B) 2 (C) 3 (D) 4 (E) 5

14. Peter built a podium (as in the picture). How many cubes did he use?

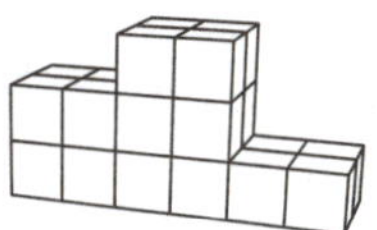

(A) 12 (B) 18 (C) 19 (D) 22 (E) 24

15. There are 5 children in a family. Kitty is 2 years older than Betty, but 2 years younger than Dannie. Teddy is 3 years older than Annie. Betty and Annie are twins. Who of them is the eldest?

(A) Annie (B) Betty (C) Dannie (D) Kitty (E) Teddy

16.    Which is next?

(A) 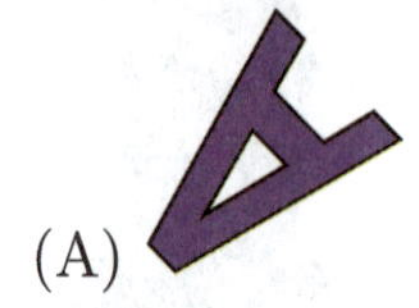 (B) (C)  (D) 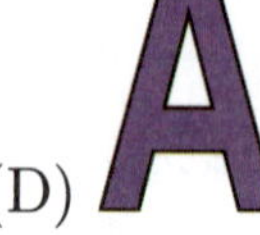 (E) 

Problems 5 points each

17. Kasia has 3 brothers and 3 sisters. How many brothers and how many sisters does her brother Mike have?

(A) 3 brothers and 3 sisters (B) 3 brothers and 4 sisters (C) 2 brothers and 3 sisters
(D) 3 brothers and 2 sisters (E) 2 brothers and 4 sisters

18. In a certain game it is possible to make the following exchanges:

Adam has 6 pears. How many strawberries will Adam have after he trades all his pears for just strawberries?

(A) 12 (B) 36 (C) 18 (D) 24 (E) 6

19. Ann has a square sheet of paper: . She cuts these pieces: 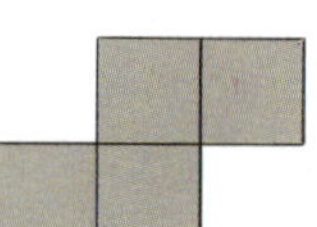 out of the sheet, as many as possible. How many pieces does she get?

(A) 1 (B) 2 (C) 3 (D) 4 (E) 5

20. Sophie makes a row of 10 houses with matchsticks. In the picture you can see the beginning of the row. How many matchsticks does Sophie need altogether?

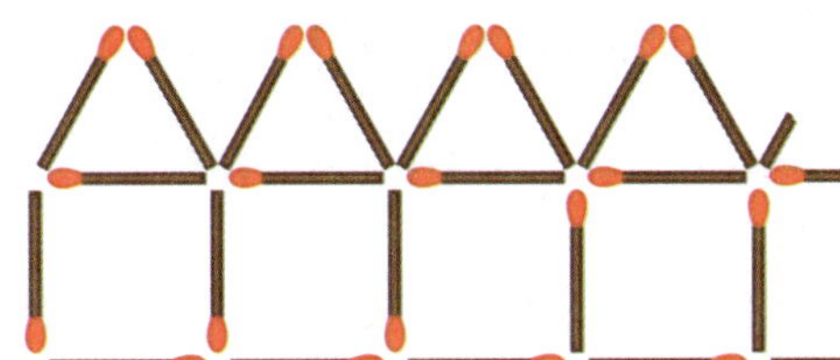

(A) 50 (B) 51 (C) 55 (D) 60 (E) 62

21.

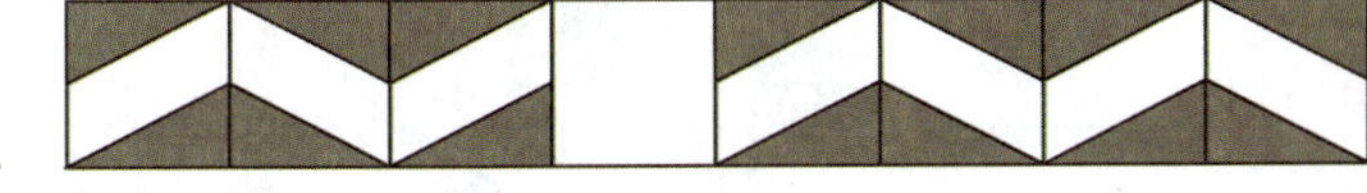

A tile falls off the wall. Caroline has three extra tiles (see the picture below).

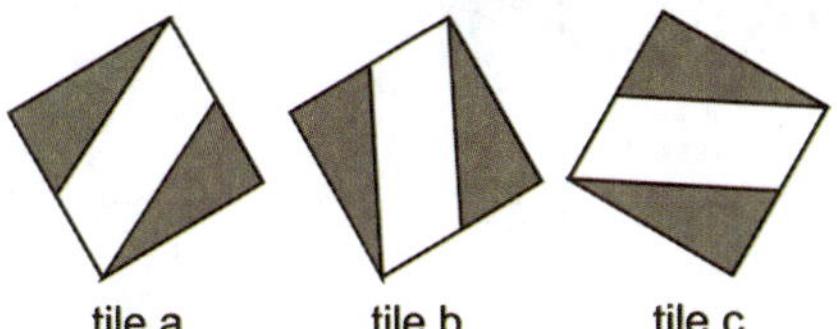

Which tiles fit the pattern?

(A) only b (B) a and b (C) b and c (D) only c
(E) all three of them fit

22. Ana has one 5-cent coin, one 10-cent coin, one 20-cent coin and one 50-cent coin. How many different values can she make with these coins?

(A) 4 (B) 7 (C) 10 (D) 15 (E) 20

23. Ania makes a large cube from 27 small white cubes. She paints all the faces of the large cube green. Then Ania removes a small cube from four corners, as shown. While the paint is still wet, she stamps each of the new faces onto a piece of paper. How many of the following stamps can Ania make?

(A) 1 (B) 2 (C) 3 (D) 4 (E) 5

24. A square box is filled with two layers of identical square pieces of chocolate. Kirill has eaten all 20 pieces in the upper layer along the walls of the box. How many pieces of chocolate are left in the box?

(A) 16 (B) 30 (C) 50 (D) 52 (E) 70

Questions from Year 2015

Problems 3 points each

1. Look closely at these four pictures.

Which figure is missing from one of the pictures?

(A) 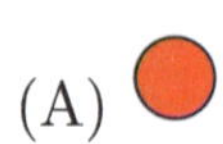 (B) 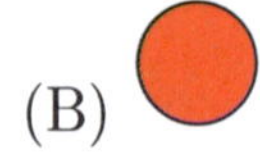 (C) 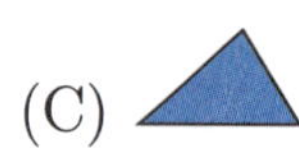 (D)  (E) 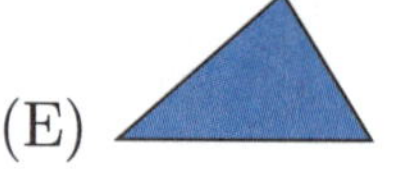

2. Find the piece missing from the house on the right.

(A) 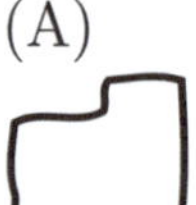 (B) 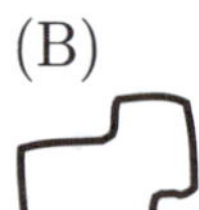 (C) 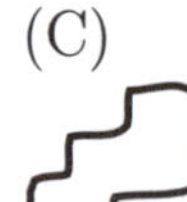 (D) 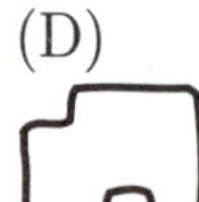 (E) 

3. There are five ladybugs shown to the left. How many spots are there on all the ladybugs together?

(A) 17 (B) 18 (C) 19 (D) 20 (E) 21

4. Which of the following pictures can be rotated so that it will be the same as the picture shown on the right?

(A)  (B) 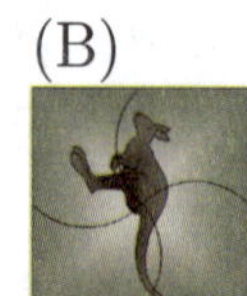 (C) 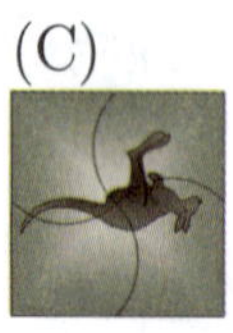 (D) 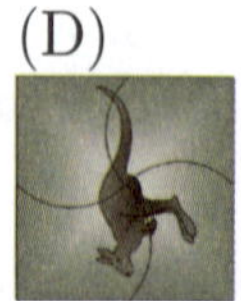 (E) 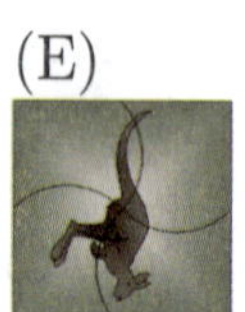

5. What does the tower shown to the right look like from above?

(A)  (B) 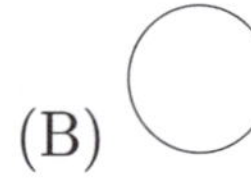 (C) 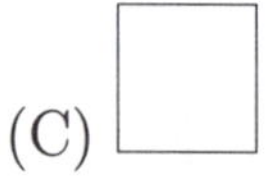 (D) 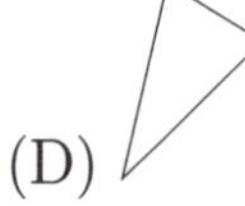 (E) 

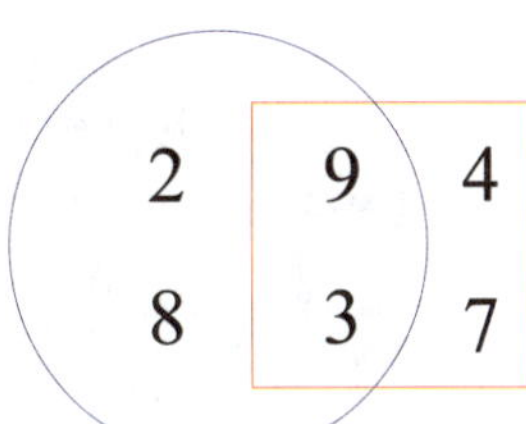

6. The picture to the left shows six numbers. What is the sum of the numbers outside the square?

(A) 12 (B) 11 (C) 23 (D) 33 (E) 10

7. Half of a movie lasts half an hour. How long does the whole movie last?

(A) 15 minutes (B) half an hour (C) 1 hour
(D) 2 hours (E) 40 minutes

8. Eric has 10 identical metal strips.

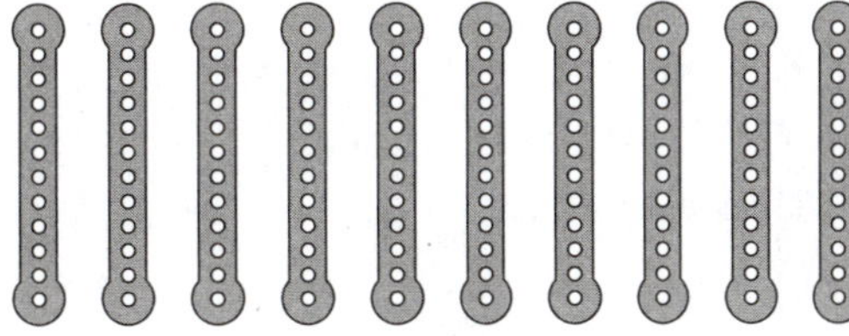

He used screws to connect pairs of them together into five long strips.

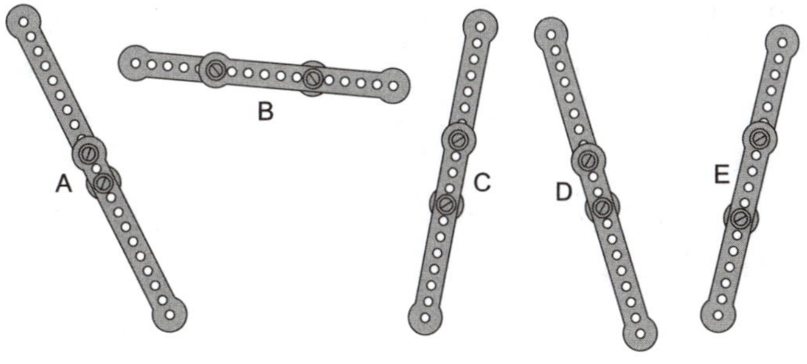

Which strip is the shortest?

(A) A (B) B (C) C (D) D (E) E

Problems 4 points each

9. There are 11 flags on a straight race track. The first flag is at the start, and the last flag is at the finish. The distance between each flag is 4 meters. How long is the track?

(A) 12 meters (B) 24 meters (C) 36 meters (D) 40 meters (E) 44 meters

10. Marko has 9 pieces of candy and Tomo has 17 pieces of candy. How many pieces of candy does Tomo need to give to Marko so that each boy has the same number of pieces of candy?

(A) 2 (B) 3 (C) 4 (D) 5 (E) 6

11. Martha built six towers using gray cubes and white cubes as shown in the picture to the right. She made each tower using five cubes. Cubes of the same color do not touch. How many white cubes did she use?

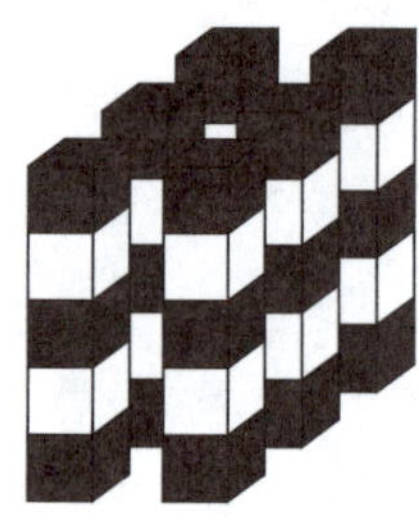

(A) 10 (B) 11 (C) 12 (D) 18 (E) 30

12. When written as 5/5/2015, the date May fifth, 2015, has three 5's. The next earliest date that will have three 5's is:

(A) May tenth, 2015 (B) April twenty-fifth, 2015 (C) May twenty-fifth, 2015
(D) January fifth, 2055 (E) May fifteenth, 2015

13. Emil placed the numbers 1, 2, 3, 4, and 5 correctly in the boxes in the diagram on the right. What number did he place in the box with the question mark?

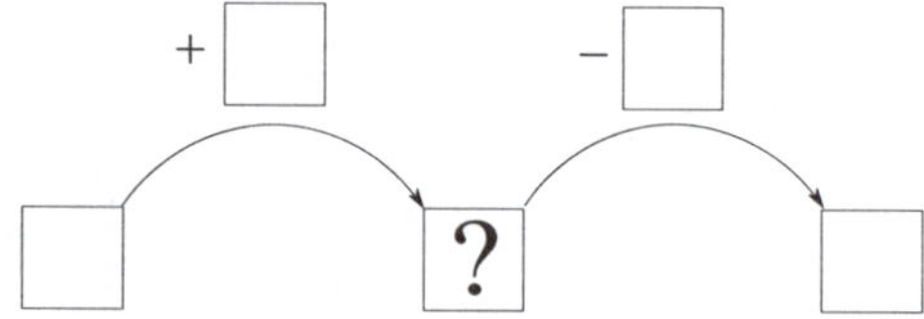

(A) 1 (B) 2 (C) 3 (D) 4 (E) 5

14. Vera invited 13 guests to her birthday party. She had 2 pizzas, and each of them was cut into 8 slices. Each person at the party ate one slice of pizza. How many slices of pizza were left over?

(A) 5 (B) 4 (C) 3 (D) 2 (E) 1

15. Don has two identical bricks (see picture to the right). Which figure can he not build using these two bricks?

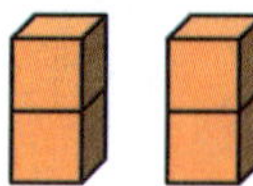

(A) 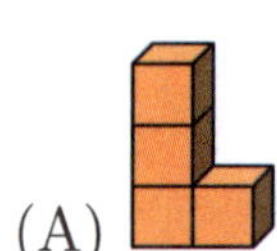 (B) 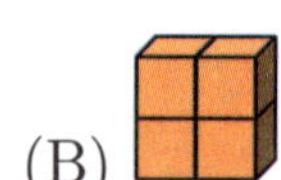 (C) 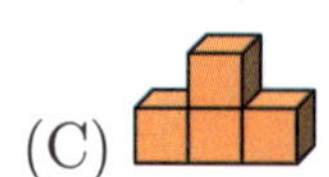 (D) 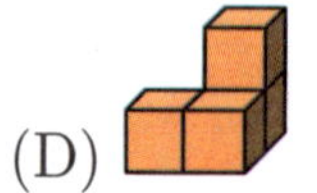 (E) 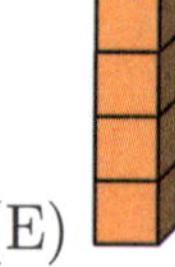

16. Which piece is missing from the puzzle to the right?

(A) 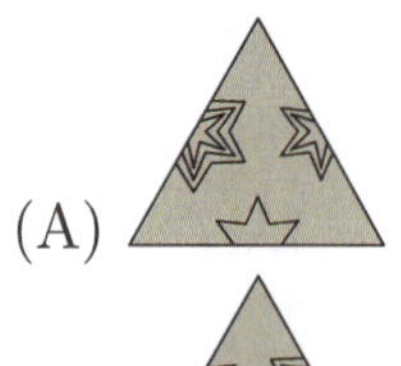 (B) 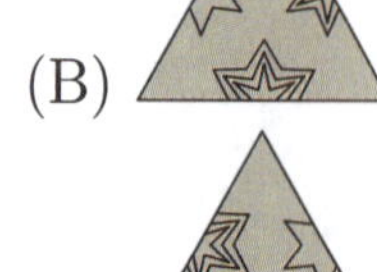 (C) 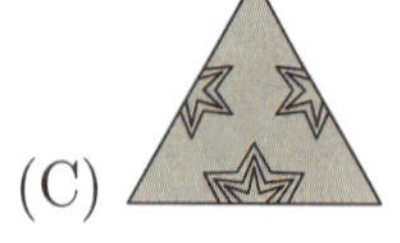

(D) 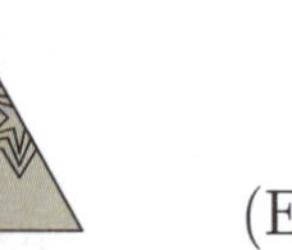 (E) 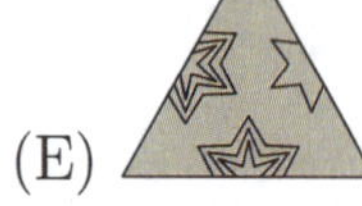

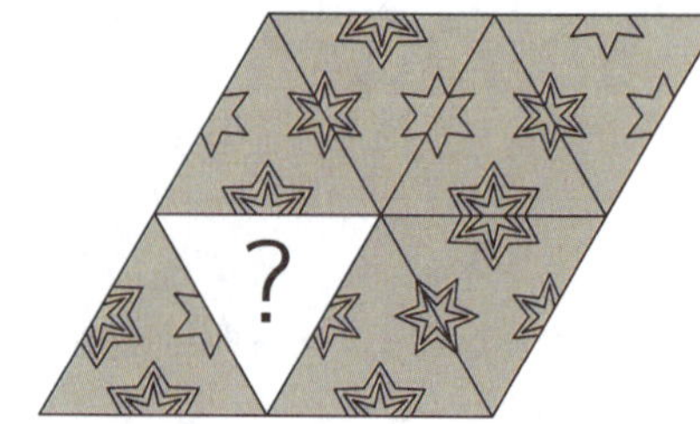

Problems 5 points each

17. In one jump, Jake the Kangaroo jumps from one circle to the neighboring circle along a line, as shown in the picture to the right. He cannot jump into any circle more than once. He starts at circle S and needs to make exactly 4 jumps to get to circle F. In how many different ways can Jake do this?

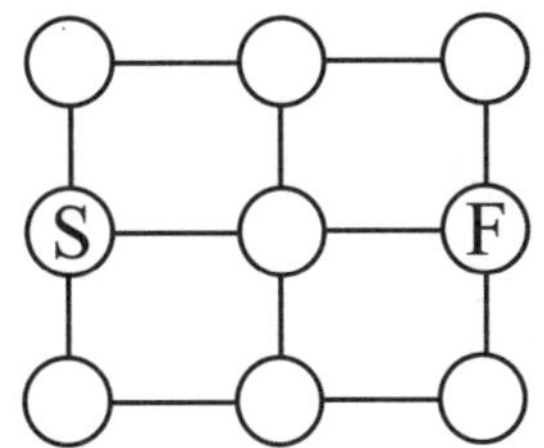

(A) 3 (B) 4 (C) 5 (D) 6 (E) 7

18. A ship was attacked by pirates. One by one the pirates climbed a rope to get to the ship. The pirate captain was the eighth pirate to climb, and there were as many pirates in front of him as behind him. How many pirates climbed the rope?

(A) 7 (B) 8 (C) 12 (D) 15 (E) 16

19. For 3 days Joy the cat was catching mice. Each day Joy caught 2 mice more than the previous day. On the third day Joy caught twice as many mice as on the first day. In total, how many mice did Joy catch during the three days?

(A) 12 (B) 15 (C) 18 (D) 20 (E) 24

20. The numbers 3, 5, 7, 8, and 9 were written in the squares of the cross (see figure to the right) so that the sum of the numbers in the row is equal to the sum of the numbers in the column. Which number was written in the central square?

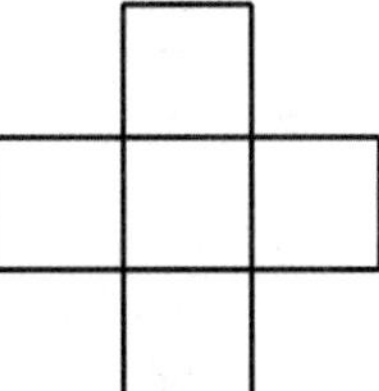

(A) 3 (B) 5 (C) 7 (D) 8 (E) 9

21. My grandmother has a dog named Atos, as well as some ducks, hens, and geese. She has 40 animals altogether. She has four times as many geese as ducks, and Atos together with the hens makes up one half of all her animals. My grandmother has:

(A) 19 hens and 5 ducks (B) 20 hens and 4 ducks (C) 19 hens and 15 geese
(D) 19 hens and 16 geese (E) 20 hens and 16 geese

22. One of the six stickers shown below was placed on each of the six faces of a die.

The next picture shows the die in two positions.

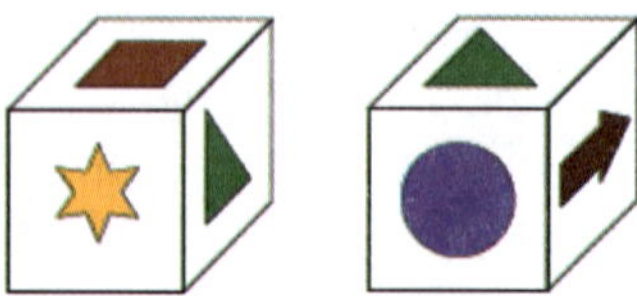

Which picture is on the face opposite the face with the kangaroo sticker?

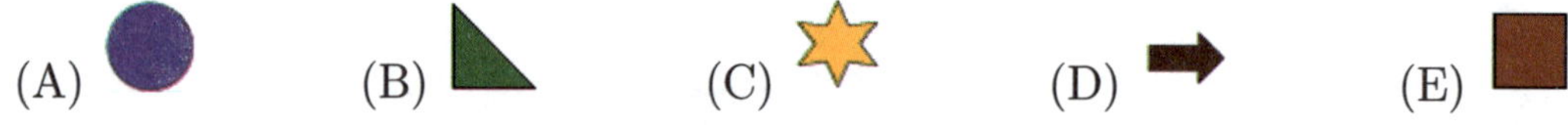

23. Sylvia, Tara, Una, and Wanda went out for dessert. They stood in line one after another. Each one of them ordered one of the following desserts: ice cream, waffle, bun, and cake, and each one ordered a different item. We know that:

- The first girl did not buy the ice cream or the waffle.
- Una was not last in line, and she bought the cake.
- Sylvia, who was standing behind Tara and in front of Una, did not buy the waffle.

Which of the following is true?

(A) Wanda was first in line.

(B) Tara bought the ice cream.

(C) Wanda bought the waffle.

(D) Una was second in line.

(E) Sylvia bought the bun.

24. We left for a summer camp yesterday at 4:32 p.m. and got to our destination today at 6:11 a.m. How long did we travel?

(A) 13 hours 39 minutes (B) 14 hours 39 minutes (C) 14 hours 21 minutes
(D) 13 hours 21 minutes (E) 2 hours 21 minutes

Questions from Year 2017

Problems 3 points each

1. Who caught the fish?

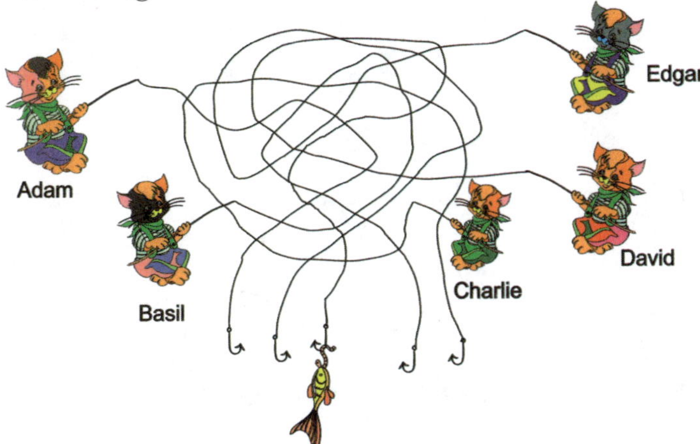

(A) Adam (B) Basil (C) Charlie (D) David (E) Edgar

2. In the picture there are stars with 5 points, stars with 6 points, and stars with 7 points. How many stars that have only 5 points are there?

(A) 2 (B) 3 (C) 4 (D) 5 (E) 9

3. The entire pie shown in the picture is divided among several children. Each child receives a piece of pie with three cherries on top. How many children are there?

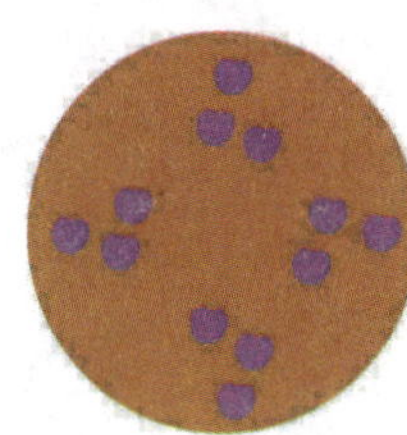

(A) 3 (B) 4 (C) 5 (D) 6 (E) 8

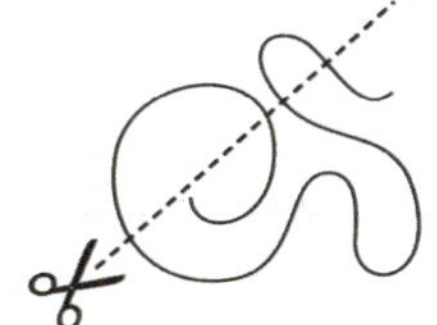

4. Into how many parts do the scissors cut the rope in the picture?

(A) 5 (B) 6 (C) 7 (D) 8 (E) 9

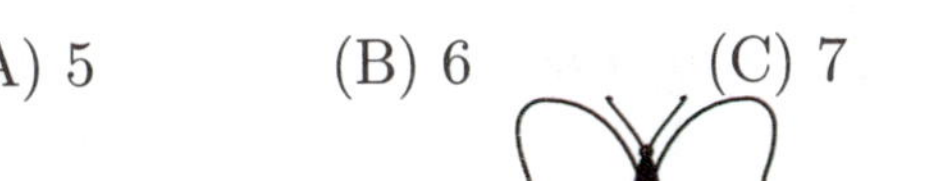

5. Ellen wants to decorate the butterfly with these stickers: . Which butterfly can she make?

(A) 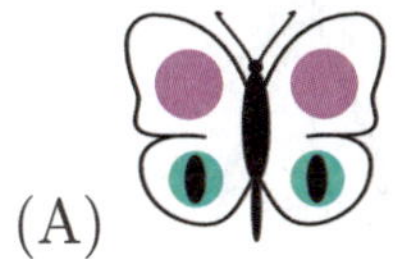 (B) 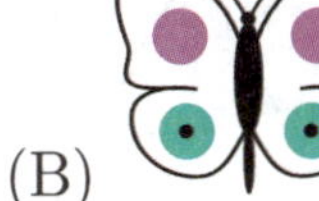 (C)  (D) 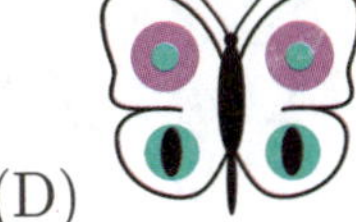 (E) 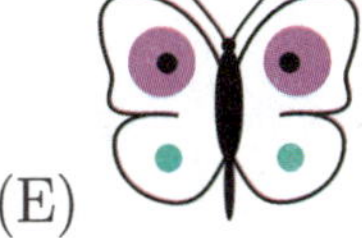

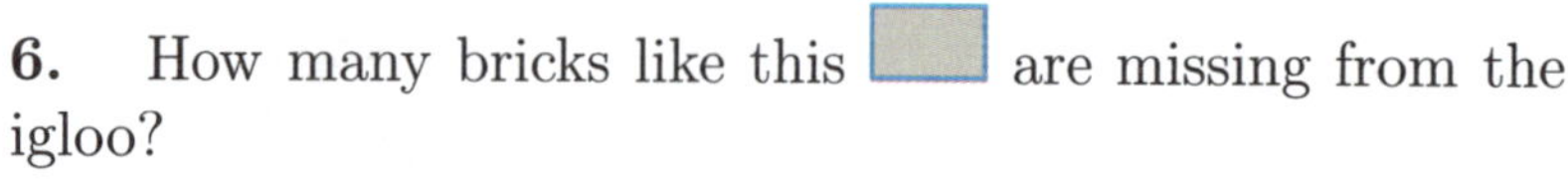

6. How many bricks like this are missing from the igloo?

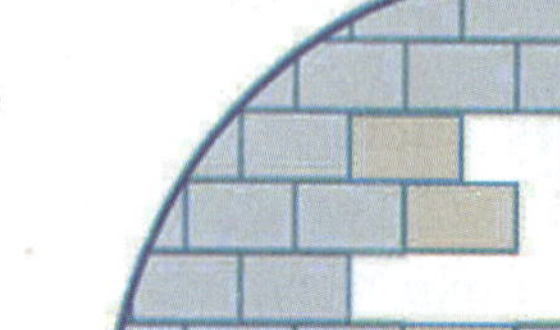

(A) 6 (B) 7 (C) 8 (D) 9 (E) 10

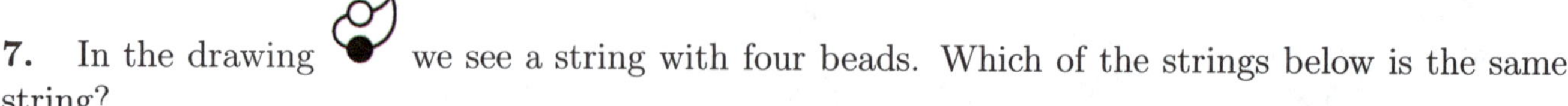

7. In the drawing we see a string with four beads. Which of the strings below is the same string?

(A) 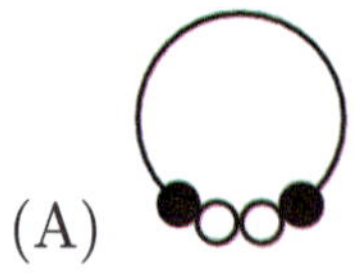 (B)  (C) 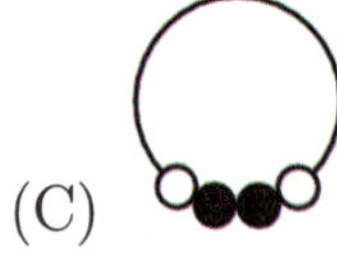 (D) 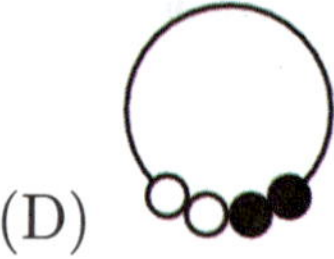 (E) 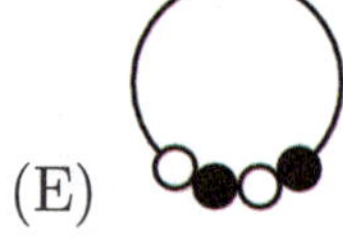

8. Four of the numbers 1, 3, 4, 5, and 7 are used, one in each square, so that the number sentence is correct. Which of the numbers is not used?

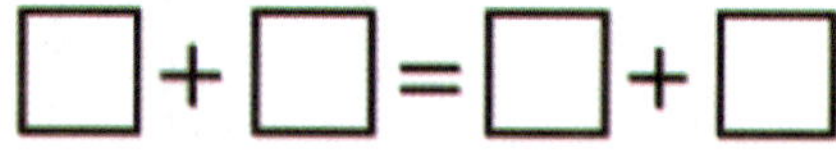

(A) 1 (B) 3 (C) 4 (D) 5 (E) 7

Problems 4 points each

9. In the country of Jewelries you can trade three sapphires for one ruby (picture 1). For one sapphire you can get two flowers (picture 2). How many flowers can you get for two rubies?

(A) 6 (B) 8 (C) 10 (D) 12 (E) 14

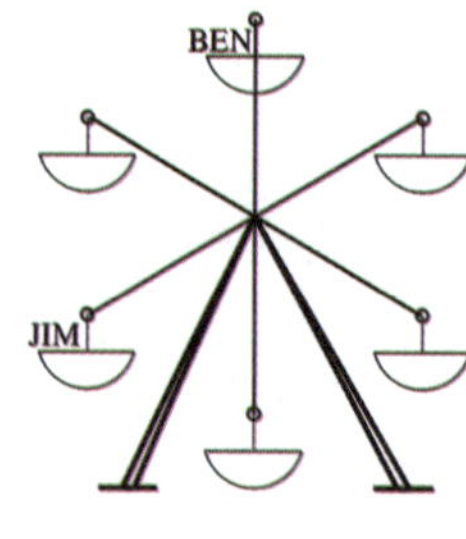

10. At one point Jim and Ben sat on the Ferris wheel as shown in the picture to the left. The Ferris wheel turned moving Ben to the place where Jim was previously. At that moment where was Jim?

(A) 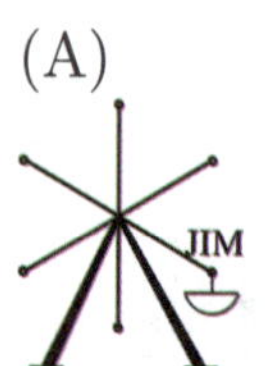

(B) 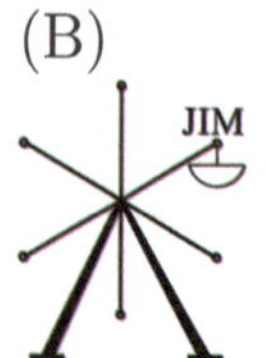

(C) 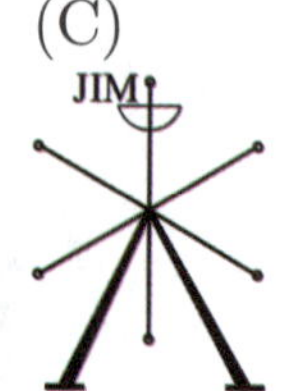

(D) 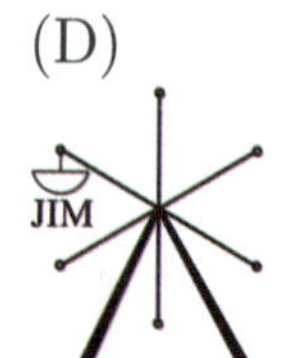

(E) 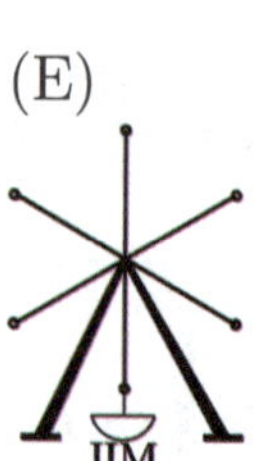

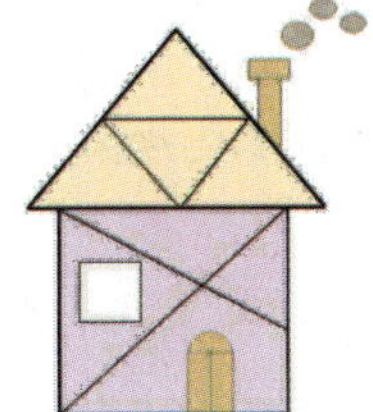

11. How many triangles are there in the picture on the right?

(A) 8 (B) 9 (C) 10 (D) 11 (E) 12

12. Alfred was rotating a shape. The first three turns are shown in the picture. He did six turns in total. How does the shape look at the end?

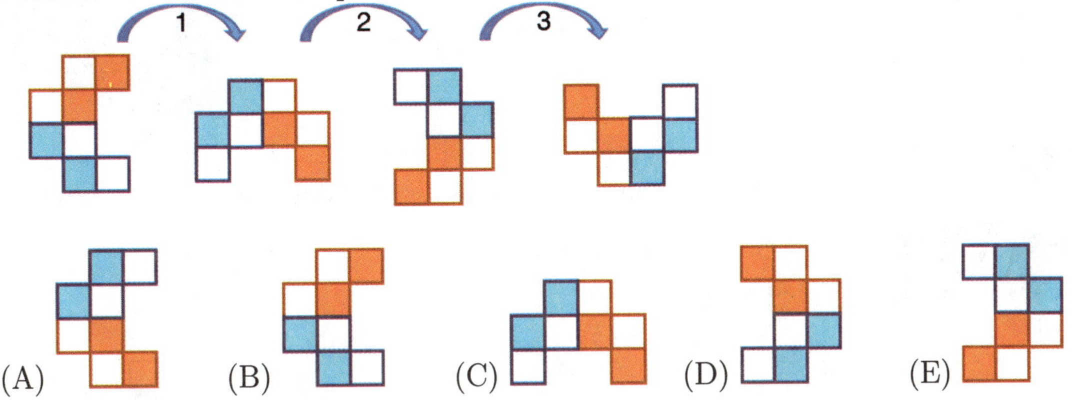

(A) 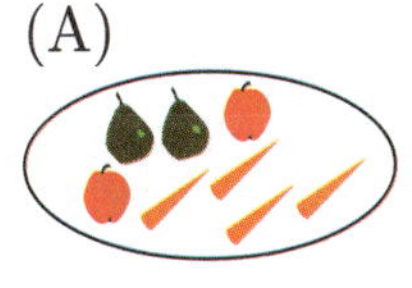 (B)  (C)  (D) 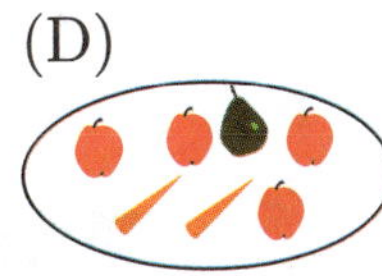 (E) 

13. In which picture are there twice as many apples as carrots and twice as many carrots as pears?

(A) (B) (C) (D) (E)

14. Brian and William are standing in line. Brian knows that there are 7 people in front of him. William knows that there is a total of 11 people in the line. If Brian is just in front of William, how many of the people in the line are behind William?

(A) 2 (B) 3 (C) 4 (D) 5 (E) 6

15. The ancient Romans used Roman numerals. We still use them today. Here are some examples: I=1, II=2, V=5, IX = 9, X=10, XI=11, XX=20. This year, 2017, we celebrate Math Kangaroo number XX. What year was Math Kangaroo number XV?

(A) 2010 (B) 2011 (C) 2012 (D) 2013 (E) 2014

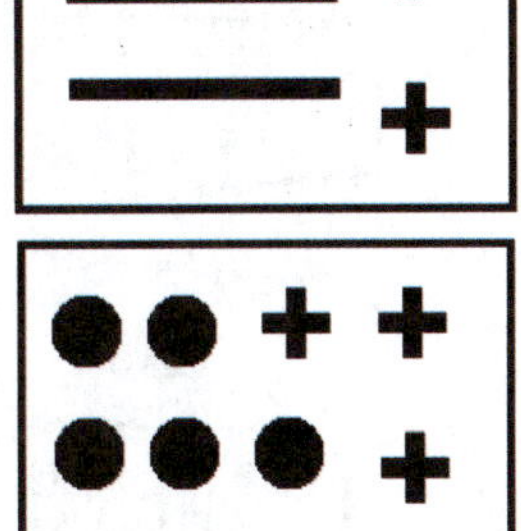

16. Liz is making crowns like this , using the following stickers: , and . There are two types of sticker sheets, shown on the right. If she wants to make 3 crowns, what is the smallest number of sheets that she will use?

(A) 3 (B) 4 (C) 5 (D) 6 (E) 7

Problems 5 points each

17. In the table, the correct additions were performed in the squares according to the pattern shown. What number should replace the question mark?

(A) 10 (B) 11 (C) 12 (D) 13 (E) 15

18. In Old McDonald's Barn there is one horse, two cows, and three pigs. How many more cows does Old McDonald's Barn need so that the number of all the animals is twice the number of cows?

(A) 0 (B) 1 (C) 2 (D) 3 (E) 4

19. Sepehr has two paper cutouts. He colored one side of each cutout like this: . Which shape can he make using both pieces?

(A) 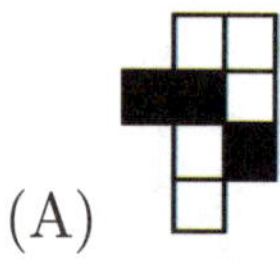 (B) 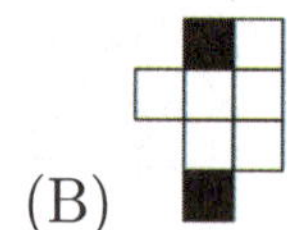 (C) 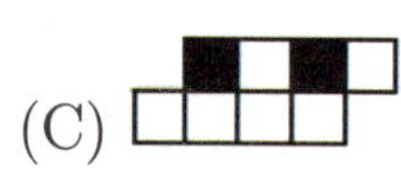 (D) 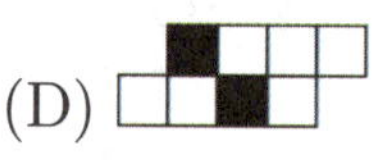 (E) 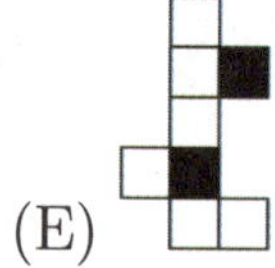

20. A certain kangaroo makes 10 jumps in 1 minute and rests 3 minutes after. Then he again makes 10 jumps in 1 minute and rests 3 minutes, and so on. What is the least number of minutes that he needs to make 30 jumps?

(A) 4 (B) 5 (C) 7 (D) 8 (E) 9

21. Which stamp was used to get the picture shown to the right?

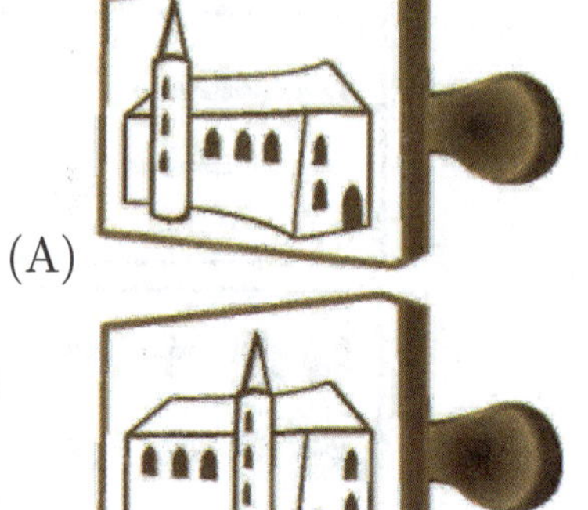

(A)

(B)

(C)

(D)

(E)

22. Each of the 4 keys fits only one of the 4 locks and the numbers on the keys refer to the letters on the locks.

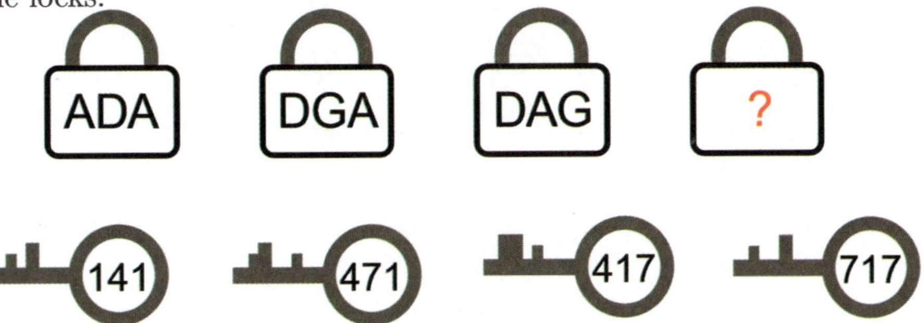

What letters are written on the lock with the question mark?

(A) GDA (B) ADG (C) GAD (D) GAG (E) DAD

23. Ann put six toys in one shelf with six cubbies, like this: 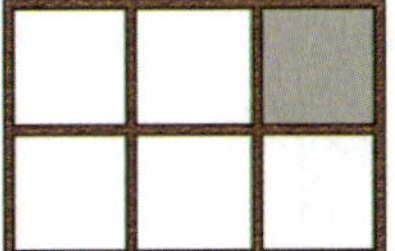.
When you look at the shelf, you see that:

-  is between  and 
-  is right above
-  is to the left of  and to the right of .

Which toy is in the gray cubby?

(A)  (B) (C)  (D)  (E) 

24. In a stack of three cards with holes, the top of each card is white and the bottom is gray. Basil threaded these cards on a rope (see picture on the right). Which of the following can he obtain without untying the rope?

(A) 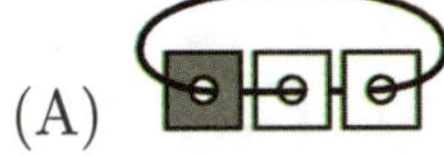 (B) 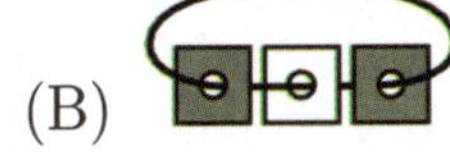 (C) 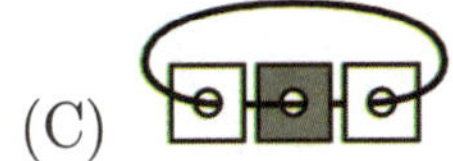

(D)  (E) 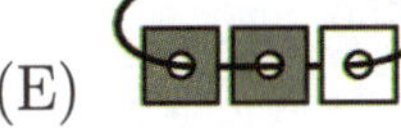

Questions from Year 2019

Problems 3 points each

1. Which cloud contains only numbers less than 7?

(A) 

(B) 

(C) 

(D) 

(E) 

2. Which figure shows a part of this necklace?

(A)  (B)  (C)  (D)  (E) 

3. Together, Mom Kangaroo and her son Jumper weigh 60 kilograms. Mom Kangaroo alone weighs 52 kilograms. How much does Jumper weigh?

(A) 2 kilograms (B) 4 kilograms (C) 8 kilograms
(D) 30 kilograms (E) 46 kilograms

4. Karen cuts one piece out of this grid: 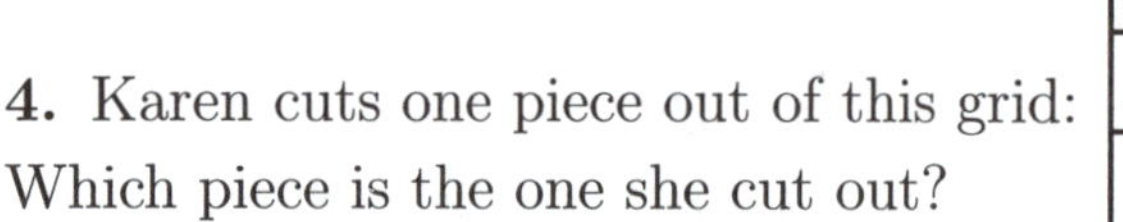.
Which piece is the one she cut out?

(A)  (B)  (C)  (D)  (E) 

5. At the entrance of the zoo there are 12 children in line. Lucy is the 7th from the front and Sam is the second from the back. How many children are there between Lucy and Sam in the line?

(A) 2 (B) 3 (C) 4 (D) 5 (E) 6

6. Jorge pairs his socks so that the numbers match. How many pairs can he make?

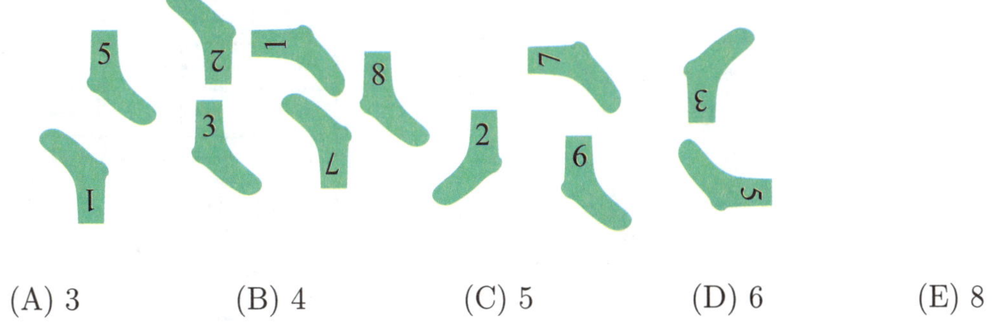

(A) 3 (B) 4 (C) 5 (D) 6 (E) 8

7. Maya the Bee was collecting pollen from all of the flowers that are inside the rectangle but outside the triangle. From how many flowers did she collect pollen?

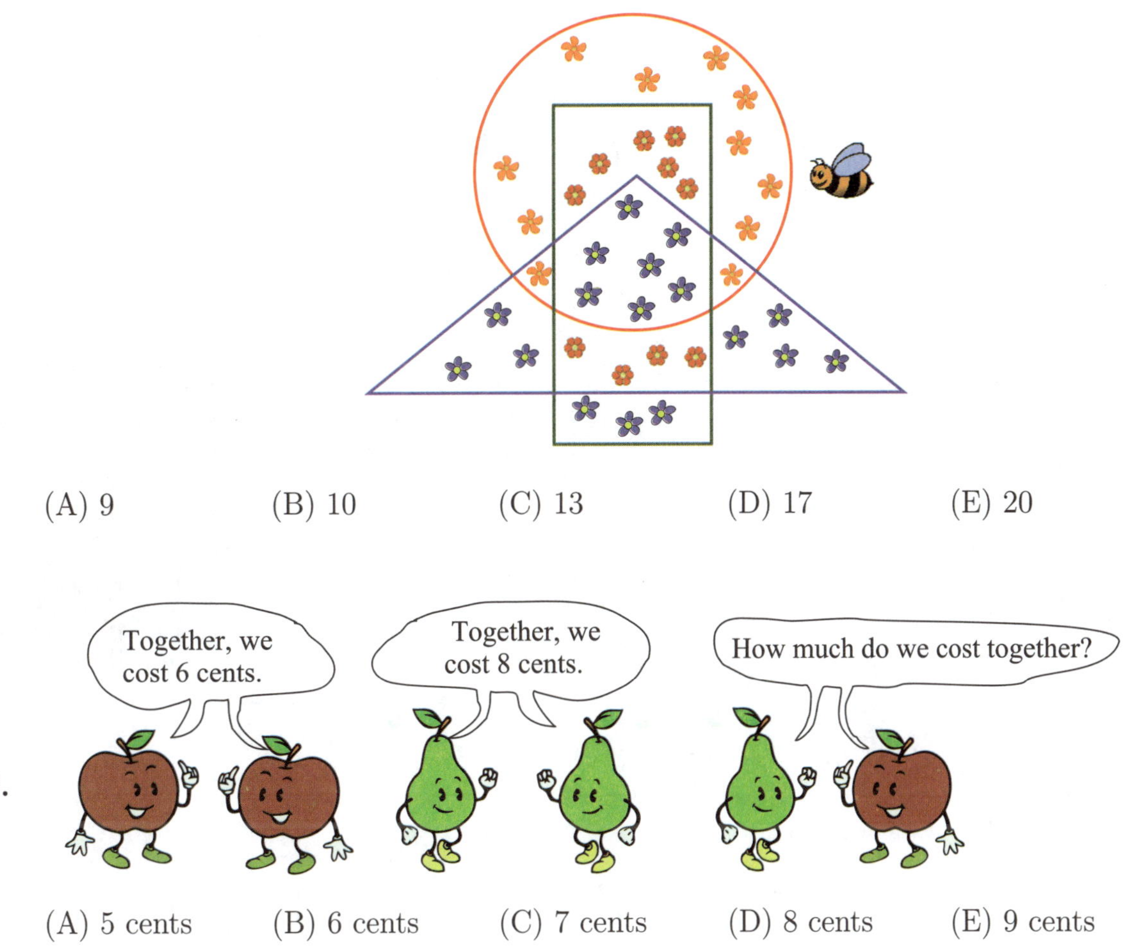

(A) 9 (B) 10 (C) 13 (D) 17 (E) 20

8.

(A) 5 cents (B) 6 cents (C) 7 cents (D) 8 cents (E) 9 cents

Problems 4 points each

9. You have to close two of the five gates so that the mouse cannot reach the cheese. Which gates should you close?

(A) 1 and 2 (B) 2 and 3 (C) 3 and 4
(D) 3 and 5 (E) 4 and 5

10. Patricia folds a sheet of paper twice and then cuts it, as shown. How many pieces of paper does she end up with?

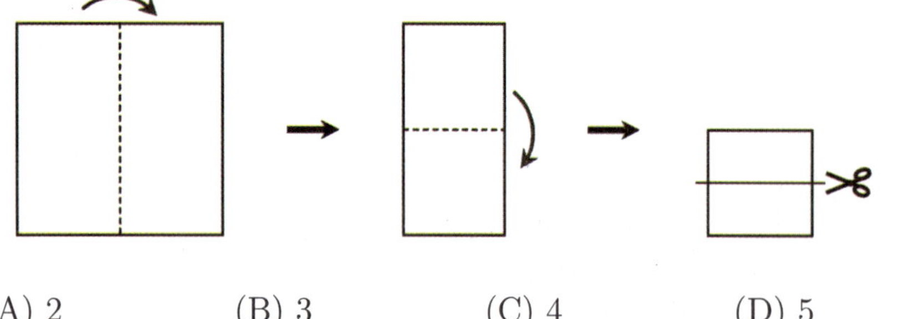

(A) 2 (B) 3 (C) 4 (D) 5 (E) 6

11. Five square cards are stacked on a table, as shown. The cards are removed one by one from the top of the stack. In what order are the cards removed?

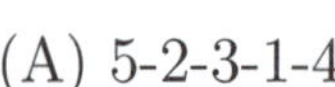

(A) 5-2-3-1-4 (B) 5-2-3-4-1 (C) 4-5-2-3-1
(D) 5-3-2-1-4 (E) 1-2-3-4-5

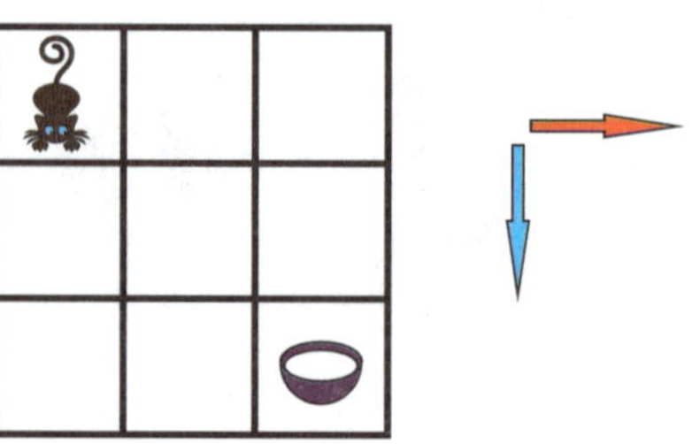

12. A cat and a bowl of milk are in opposite corners of the board. The cat can only move as shown by the arrows. In how many ways can the cat reach the milk?

(A) 2 (B) 3 (C) 4 (D) 5 (E) 6

13. Four strips are woven into a pattern, as shown:  .
What do you see when you look at it from the other side?

(A) 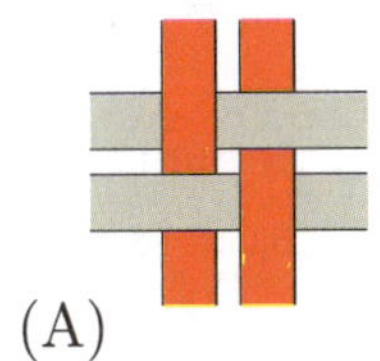 (B) 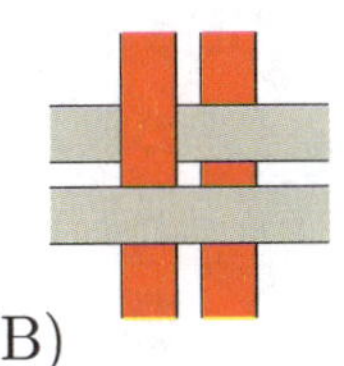 (C)  (D)  (E) 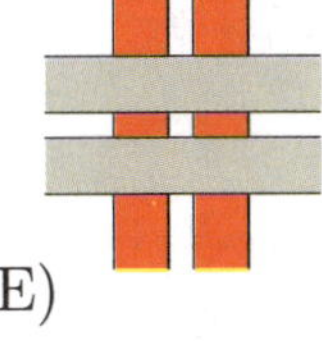

14. Each of the shapes shown is made by gluing together four cubes of the same size. The shapes will be painted. Which shape has the smallest area to be painted?

(A) 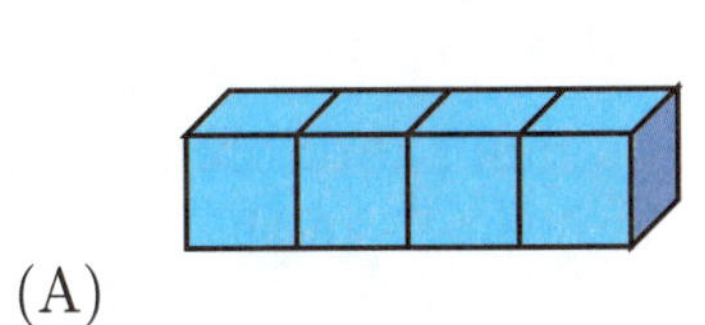 (B)  (C) 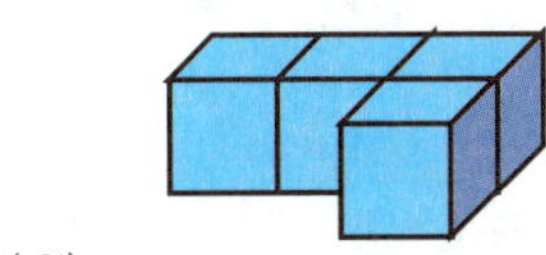

(D) 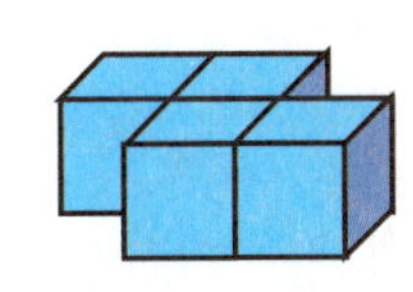 (E) 

15. A floor is covered with identical rectangular tiles as shown. The shorter side of each tile is 1 m. What is the length of the side with the question mark?

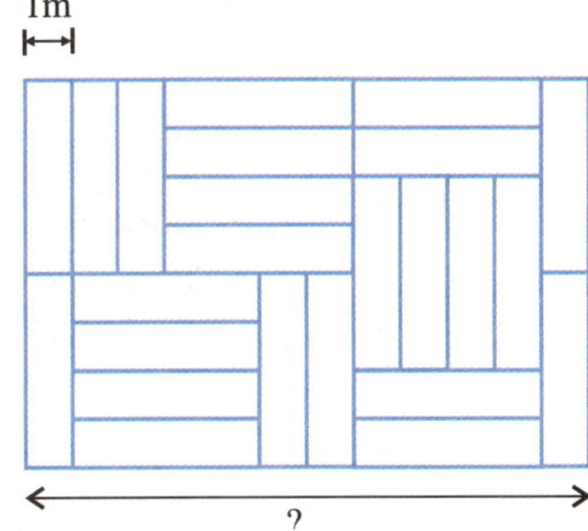

(A) 6 m (B) 8 m (C) 10 m (D) 11 m (E) 12 m

16. A train from Kang station to Aroo station leaves at 6:00 in the morning and passes by three other stations without stopping.

The numbers show the travel times between two stations, in hours. The train arrives at Aroo station at 11:00 at night on the same day. What is the travel time between Aroo station and the previous station?

(A) 2 hours (B) 3 hours (C) 4 hours (D) 5 hours (E) 6 hours

Problems 5 points each

17. On a farm, there are only sheep and cows. The number of sheep is 8 more than the number of cows. The number of cows is half the number of sheep. How many animals are on the farm?

(A) 16 (B) 18 (C) 20 (D) 24 (E) 28

18. A figure has been cut into these 3 pieces:
Which figure could have been cut?

(A)  (B) (C) 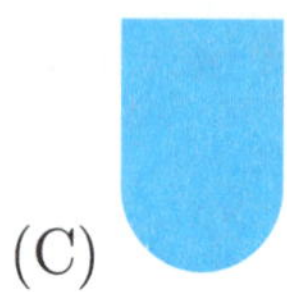 (D)  (E) 

19. There are 10 camels in a zoo. The camels are either Bactrian (with two humps) or dromedary (with one hump). In total there are 14 humps. Find the number of Bactrian camels in the zoo.

(A) 1 (B) 2 (C) 3 (D) 4 (E) 5

20. Three squirrels, Anni, Asia, and Elli, collected 7 nuts in total. Each collected a different number of nuts, but each collected at least one nut. Anni collected the least, and Asia the most. How many nuts did Elli collect?

(A) 1 (B) 2 (C) 3 (D) 4 (E) 5

21. Tim and Tom built a sandcastle and decorated it with a flag. They stuck half of the flagpole into the highest point of the castle. The upper tip of the flagpole was 80 cm above the ground and the lower tip was 20 cm above the ground. How tall was the sandcastle?

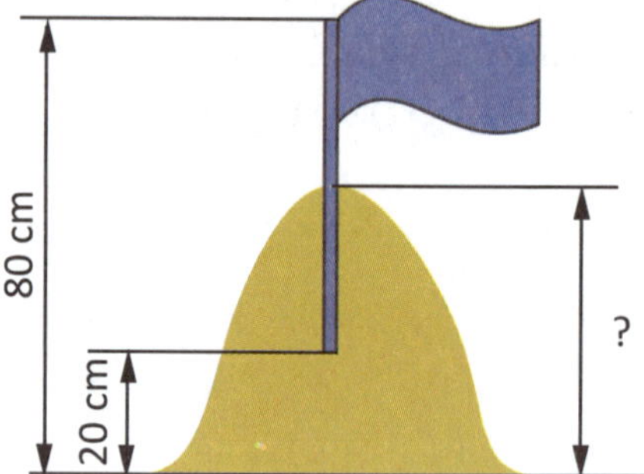

(A) 40 cm (B) 45 cm (C) 50 cm
(D) 55 cm (E) 60 cm

22. Here are nine squares: . First, Ani replaced all the black squares with white ones. Next, Bob replaced all the gray squares with black ones. Finally, Chris replaced all the white squares with gray ones. What did they get at the end?

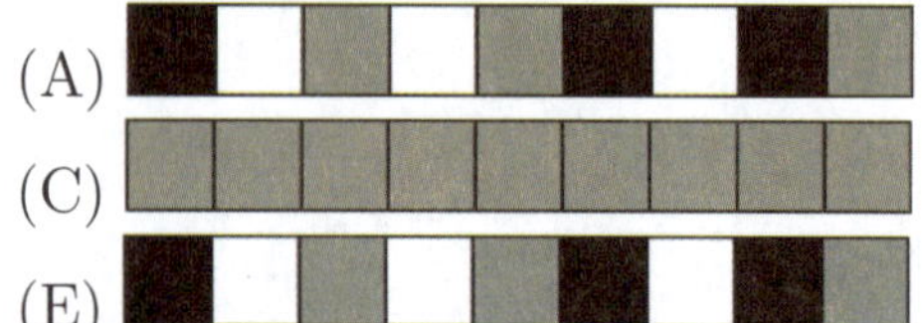
(A)
(C)
(E)

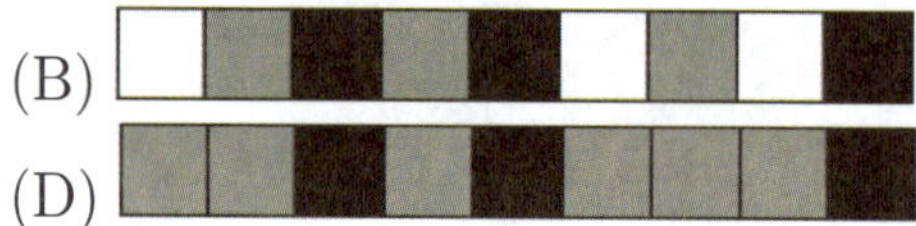
(B)
(D)

23. Peter chose a square of four cells in the table so that the sum of the four numbers inside the square is greater than 63. Which of the following numbers must be in the chosen square?

1	2	3	4	5
6	7	8	9	10
11	12	13	14	15
16	17	18	19	20

(A) 14 (B) 15 (C) 17 (D) 18 (E) 20

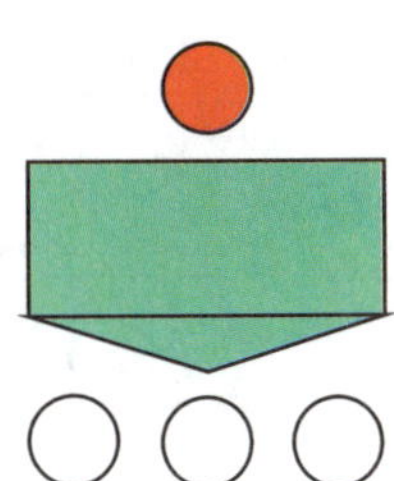

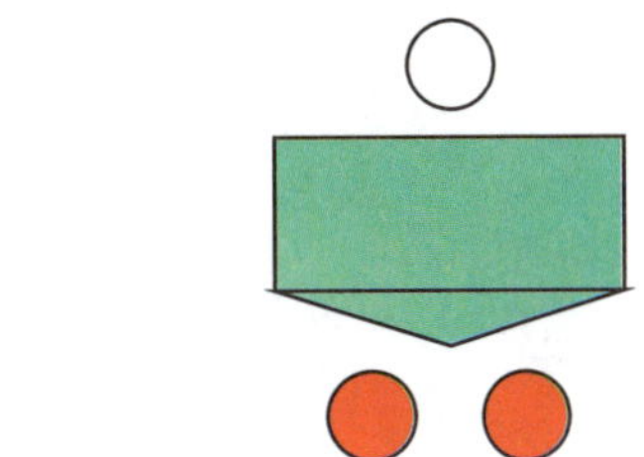

24. Amalia's machine changes one red token into three white tokens and one white token into two red tokens. Amalia has three red tokens and one white token: . She uses the machine three times. What is the smallest number of tokens she can end up with?

(A) 7 (B) 6 (C) 8 (D) 5 (E) 9

Questions from Year 2021

Problems 3 points each

1. A kangaroo laid out 3 sticks like this to make a shape. It's not allowed to break or bend the sticks. Which shape can the kangaroo make?

(A) 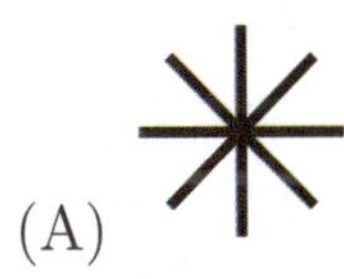 (B) (C) 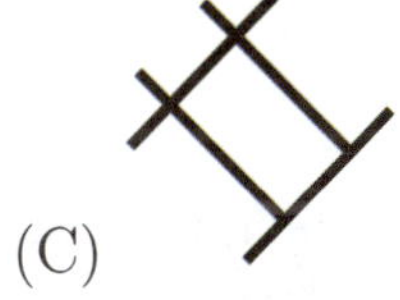 (D) 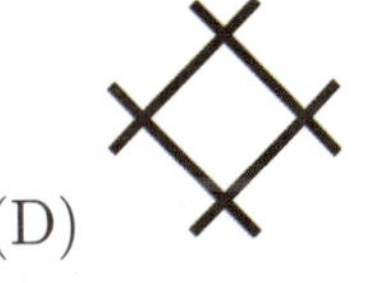 (E) 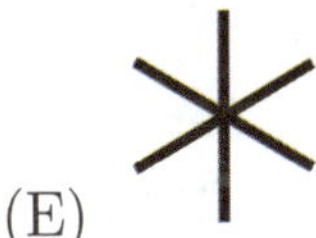

2. The picture shows 2 mushrooms. What is the difference between their heights?

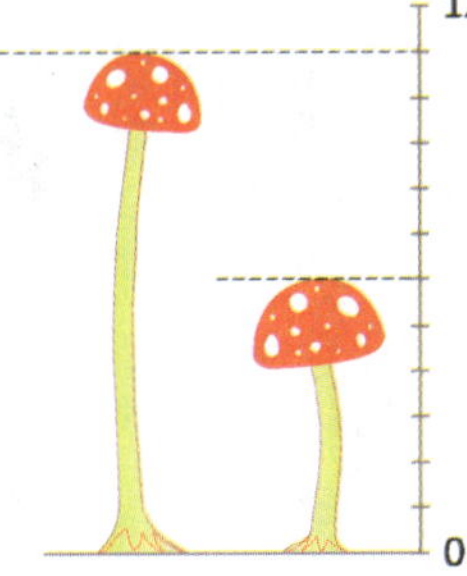

(A) 4 (B) 5 (C) 6 (D) 11 (E) 17

3. Which of the paths shown in the pictures is the longest?

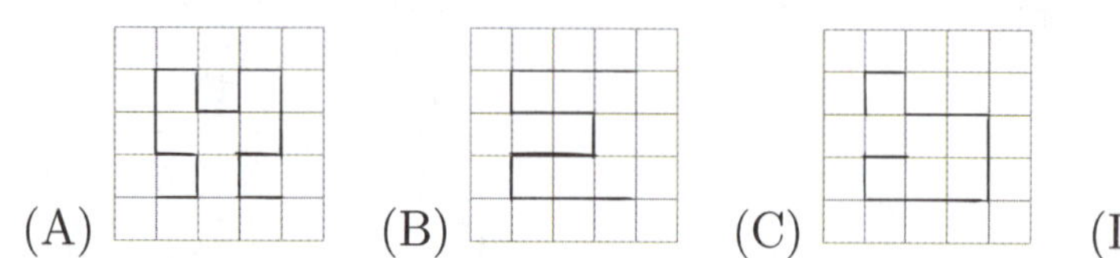

4. Four identical pieces of paper are placed as shown. Michael wants to punch a hole that goes through all four pieces. At which point should Michael punch the hole?

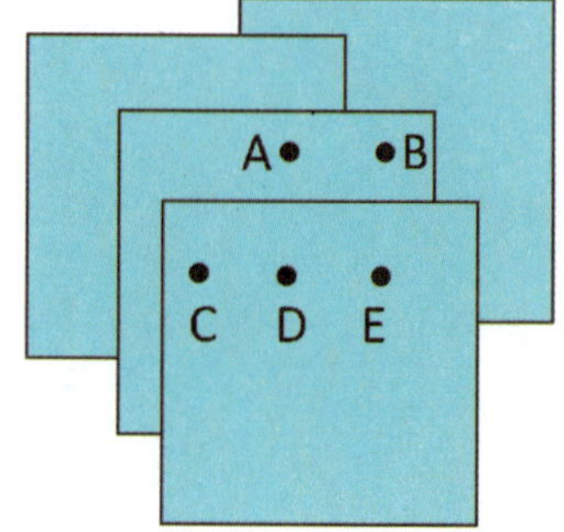

(A) A (B) B (C) C (D) D (E) E

5. Ella puts on this t-shirt and stands in front of a mirror. Which of these images does she see in the mirror?

(A) ƖƧ0Ƨ (B) Ƨ0ƧƖ (C) 0ƧƖƧ (D) 120Ƨ (E) 1Ƨ02

6. The pink tower is taller than the red tower but shorter than the green tower. The silver tower is taller than the green tower. Which tower is the tallest?

(A) pink tower (B) green tower (C) red tower (D) silver tower
(E) We don't know.

7. These children are standing in a line. Some are facing forward and others are facing backward. How many children are holding another child's hand with their right hand?

(A) 2 (B) 3 (C) 4 (D) 5 (E) 6

8. In the Kangaroo constellation, each star has a number greater than 3 and the sum of the numbers is 20. Which is the Kangaroo constellation?

(A) 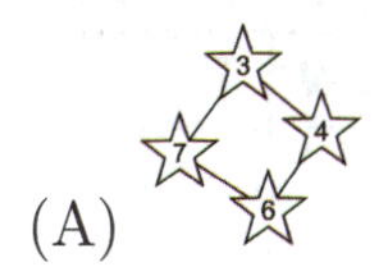

(B) 

(C) 

(D) 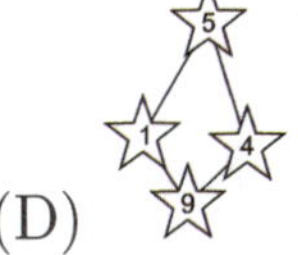

(E)

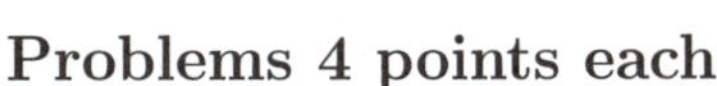

Problems 4 points each

9. Edmund cut a ribbon as shown in the picture. How many pieces of the ribbon did he end up with?

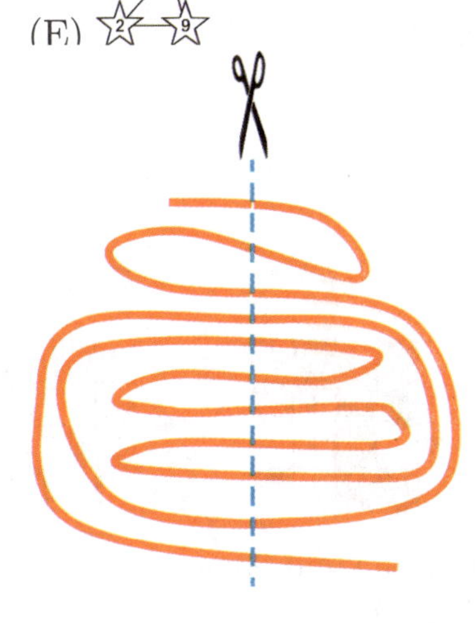

(A) 9 (B) 10 (C) 11 (D) 12 (E) 13

10. Rose the cat walks along the wall. She starts at point B and follows the direction of the arrows shown in the picture. The cat walks a total of 20 meters. Where does she end up?

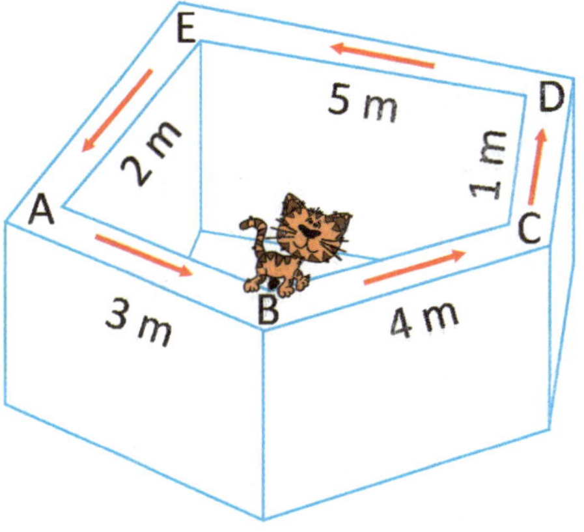

(A) A (B) B (C) C (D) D (E) E

11. Julia has two pots with flowers, as shown. She keeps the flowers exactly where they are. She buys more flowers and puts them in the pots. After that, each pot has the same number of each type of flower. What is the smallest number of flowers she needs to buy?

(A) 2 (B) 4 (C) 6 (D) 8 (E) 10

12. Tom encodes words using the board shown. For example, the word PIZZA has the code $A2\,A4\,C1\,C1\,B2$. What word did Tom encode as $B3\,B2\,C4\,D2$?

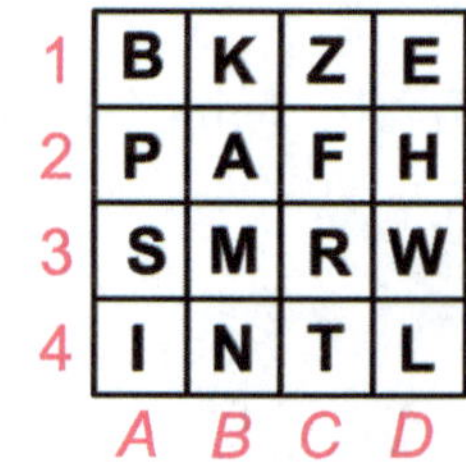

(A) MAZE (B) MASK (C) MILK (D) MATE (E) MATH

13. Which figure can be made from these two pieces?

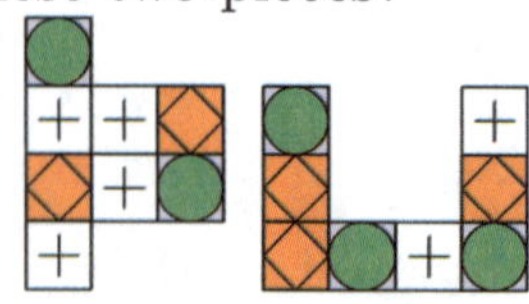

(A) (B) 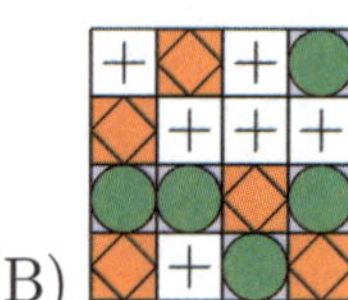 (C) 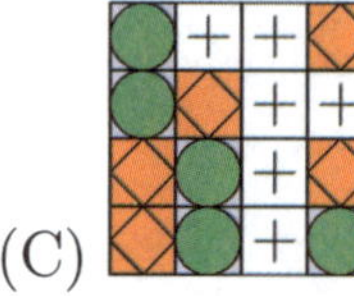 (D) 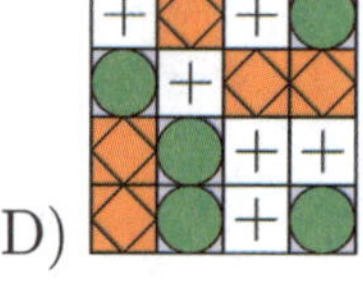 (E) 

14. Julie and Angela played "kangball," a ball game. Each goal in their game earns 2 points. Julie scored 5 goals and Angela scored 9 goals. How many more points than Julie did Angela earn?

(A) 4 (B) 6 (C) 8 (D) 10 (E) 12

15. The picture shows the five houses of five friends and their school. The school is the largest building in the picture. To go to school, Doris and Ali walk past Leo's house. Eva walks past Chole's house. Which is Eva's house?

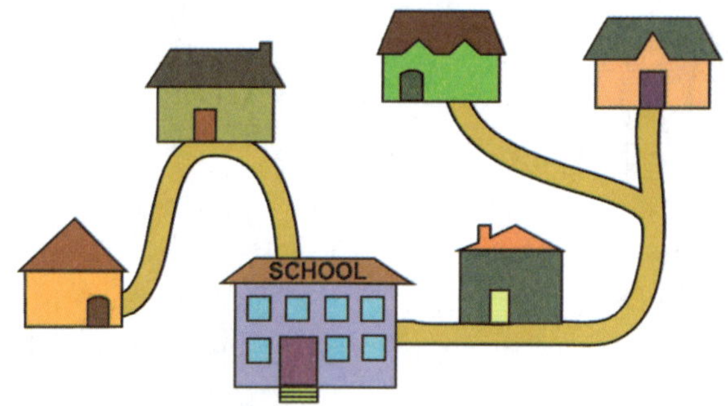

(A) (B)  (C) 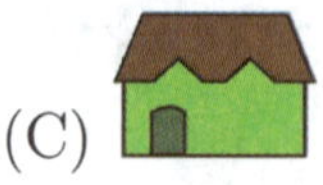 (D)  (E) 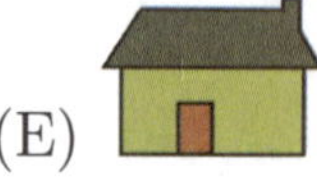

16. The kangaroo had two branches for lunch. Each branch had 10 leaves. The kangaroo ate some leaves from one branch. Then, from the second branch, it ate as many leaves as were left on the first branch. How many leaves in total were left on the two branches?

(A) 5 (B) 6 (C) 8 (D) 10 (E) 15

Problems 5 points each

17. Mara built the square shown by using four of the following five shapes. Which shape was not used?

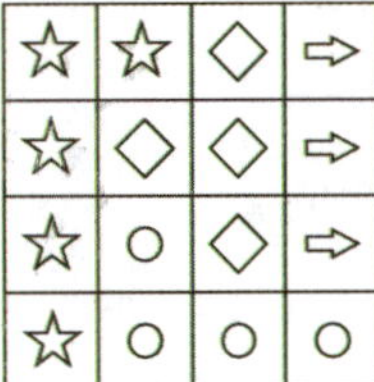

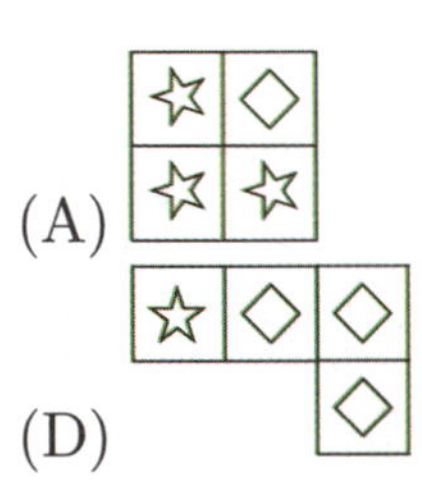

(A)

(D)

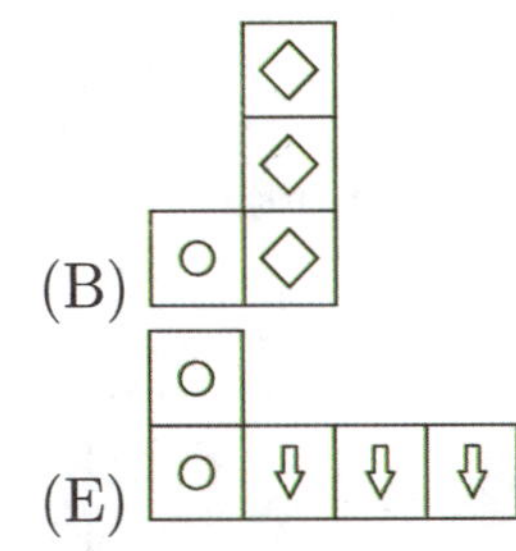

(B)

(E)

(C) 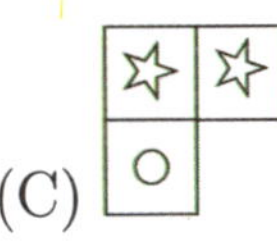

18. Every time the witch has 3 apples she turns them into 1 banana. Every time she has 3 bananas she turns them into 1 apple. What will she end up with if she starts with 4 apples and 5 bananas?

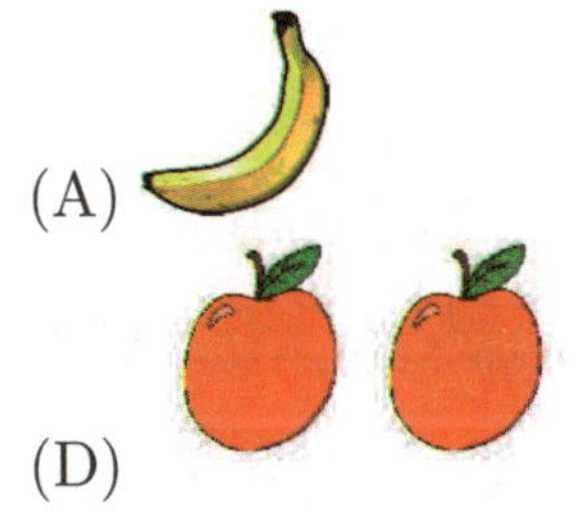

(A)

(D)

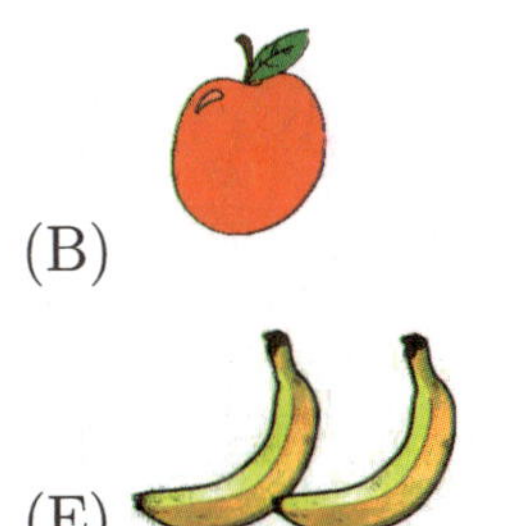

(B)

(E)

(C) 

19. The cards shown 2 3 4 5 6 are placed into two boxes. The sums of the numbers in each box are the same. Which number must be in the box with the number 4?

(A) only 3 (B) only 5 (C) only 6 (D) only 5 or 6
(E) It is impossible to know.

20. The picture shows two gears, each with a black tooth.

Where will the black teeth be after the small gear has made one full turn?

(A) 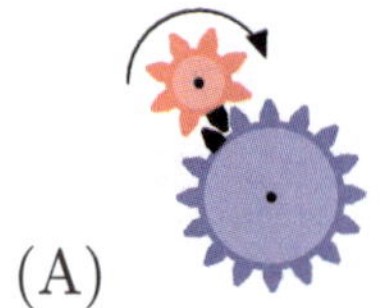 (B) 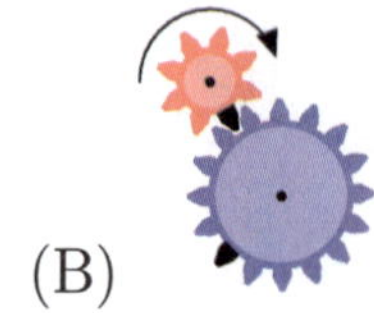 (C) 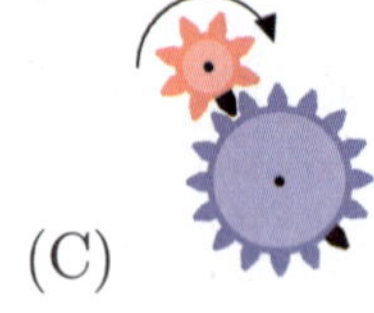 (D) 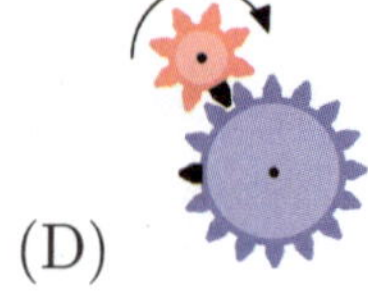 (E) 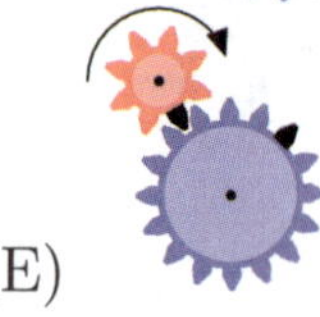

21. Three girls and two boys were dancing. They danced in pairs so that each girl danced with each boy for exactly one minute. At any time, there was only one pair on the dance floor. For how many minutes did they dance?

(A) 5 (B) 6 (C) 8 (D) 9 (E) 10

22. Each participant in a cooking contest baked one tray of cookies like the one shown.

What is the smallest number of trays of cookies needed to make the following plate?

(A) 1 (B) 2 (C) 3 (D) 4 (E) 5

23. Kangie eats only apples on Monday, Wednesday, and Friday. On Tuesdays and Thursdays he eats only mangoes. He eats either 2 apples or 3 mangoes a day. On Saturdays and Sundays he eats nothing. How many pieces of fruit does Kangie eat in two weeks?

(A) 12 (B) 16 (C) 18 (D) 20 (E) 24

24. Stan has five toys: a ball, a set of blocks, a game, a puzzle, and a car. He puts each toy on a different shelf of the bookcase. The ball is higher than the blocks and lower than the car. The game is directly above the ball. On which shelf can the puzzle not be placed?

(A) 1 (B) 2 (C) 3 (D) 4 (E) 5

Questions from Year 2023

Problems 3 points each

1. How many circles are there in the figure?

(A) 5 (B) 6 (C) 7 (D) 8 (E) 9

2. The picture shows 5 cubes viewed from the front.
What is the view from above?

(A) 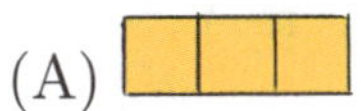 (B)  (C) 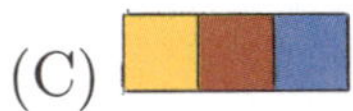 (D)  (E) 

3. Each bowl contains four balls with numbers, as shown. In which bowl is the sum of all the numbers the largest?

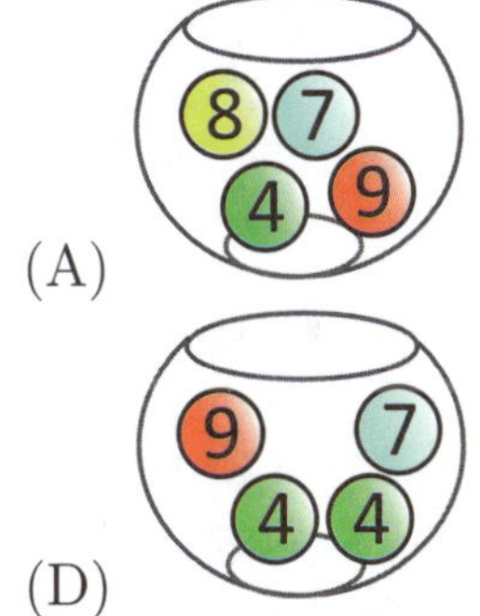

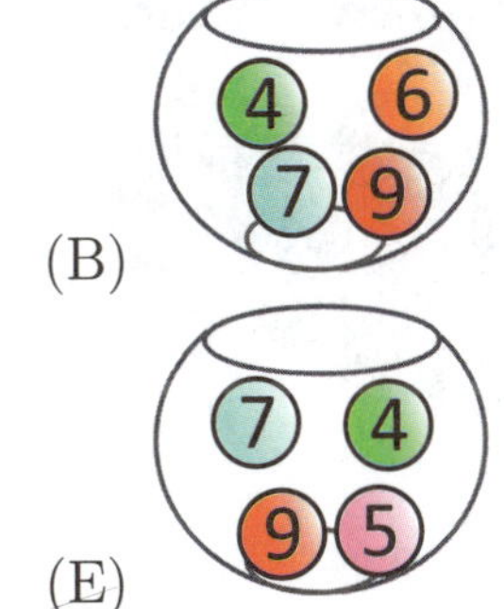

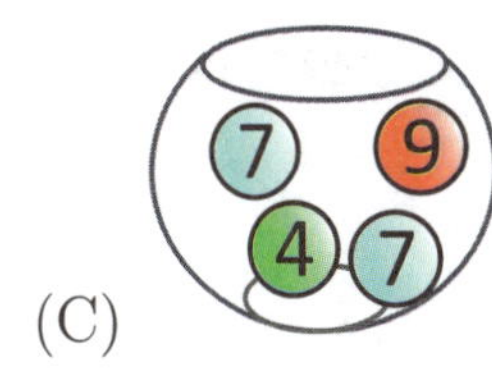

(C)

4. Mr. Beaver uses the pieces to make a kangaroo figure.

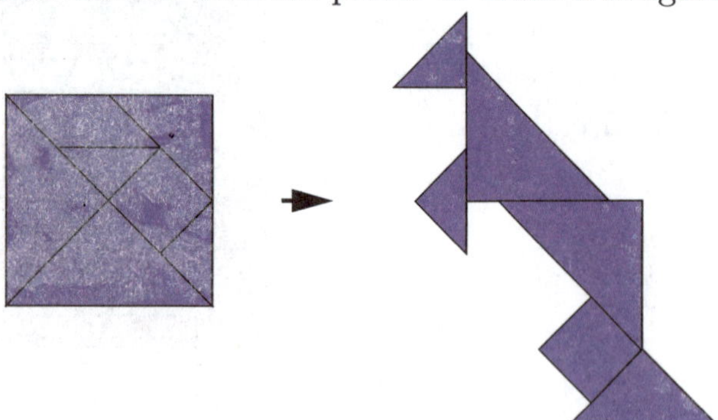

Which piece is missing?

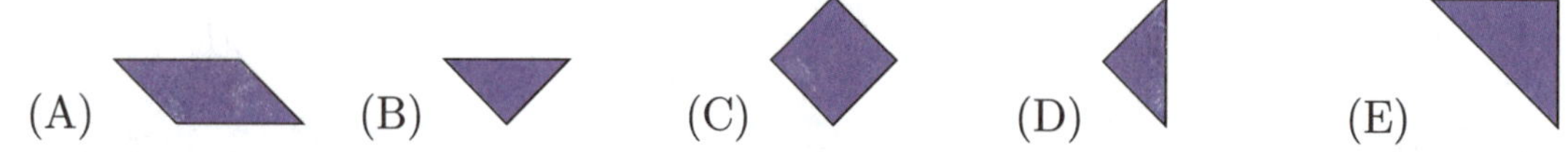

5. My boat has more than 1 circle. It also has 2 more triangles than squares. Which boat is mine?

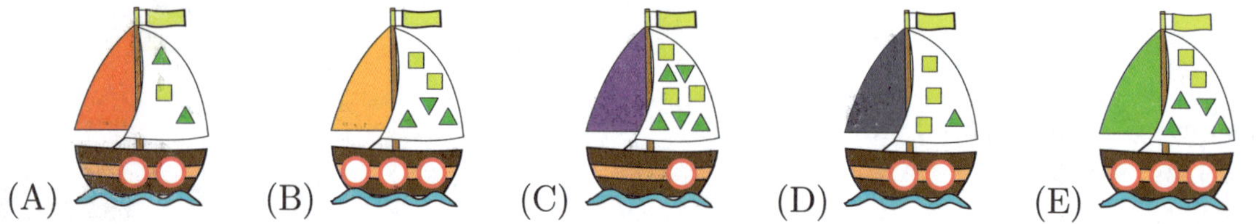

6. This is my grandfather's birthday cake. A large candle stands for 10 years and a small candle for 1 year. How old is my grandfather?

(A) 65 (B) 66 (C) 76
(D) 77 (E) 78

7. Pablo put 10 toy cars on this racetrack. How many cars are in the tunnel?

(A) 5 (B) 6 (C) 7 (D) 8 (E) 9

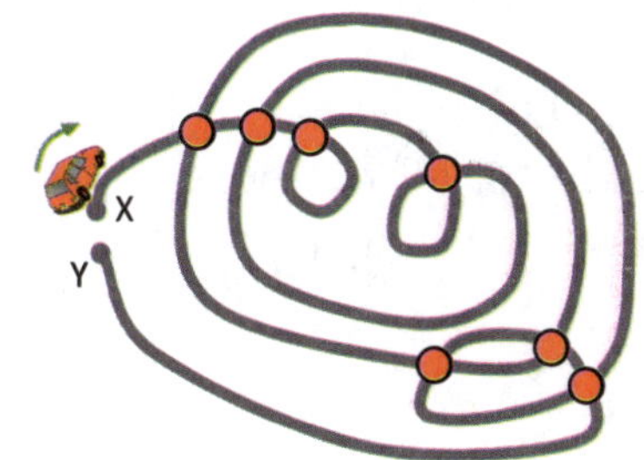

8. Steven drives from X to Y. At each crossing, he stops before going straight ahead. In total, how many times does he stop at crossings?

(A) 11 (B) 12 (C) 13 (D) 14 (E) 15

Problems 4 points each

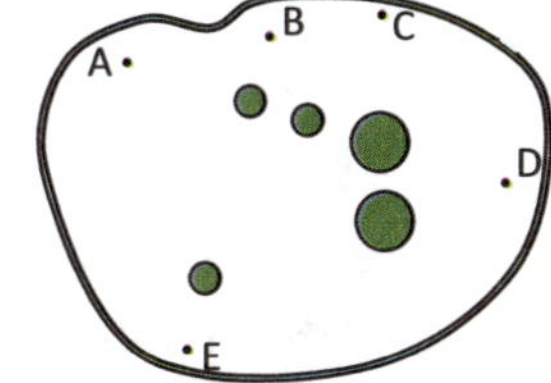

9. There are 5 trees in a park. A beaver can see only two of the trees because all the others are hidden behind other trees. At which of the marked points is the beaver standing?

(A) at A (B) at B (C) at C (D) at D (E) at E

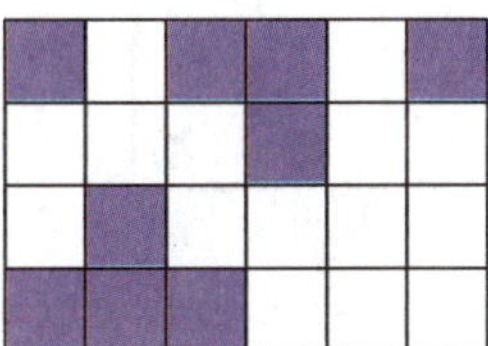

10. There are 24 squares in the picture. Aaron colored some of the squares. How many more squares does he need to color so that half of the squares are colored?

(A) 1 (B) 2 (C) 3 (D) 4 (E) 5

11. The two tokens with the question mark have the same number. The sum of the four tokens shown is 18. What is the value of one of the missing numbers?

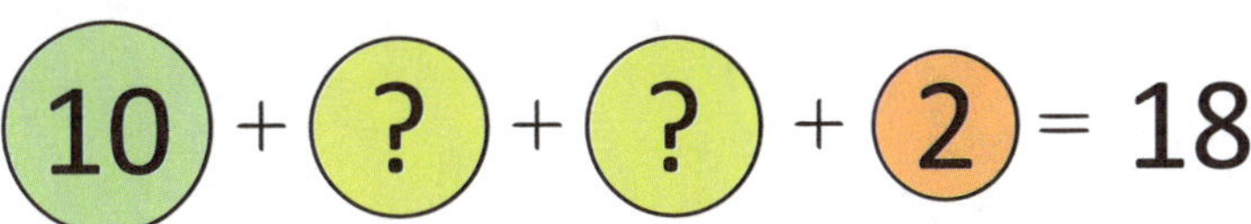

(A) 1 (B) 2 (C) 3 (D) 4 (E) 5

12. Sarah wants to finish the bee on the left according to the model on the right.

Sarah needs to use points to unlock parts of the bee. How many points does she need to use to complete the bee?

(A) 9 (B) 10 (C) 11 (D) 12 (E) 13

13. The table has 30 boxes. We paint all the boxes in row 3, all the boxes in row 6, all the boxes in column C, and all the boxes in column D. How many boxes will be not painted?

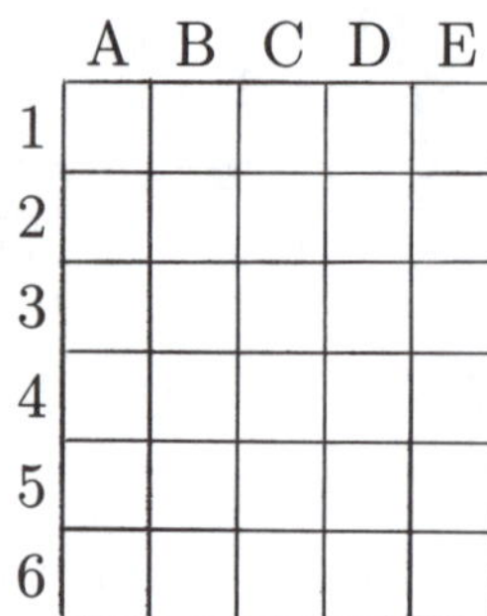

(A) 8 (B) 10 (C) 12 (D) 18 (E) 22

14. A sheet of paper is folded in half. Square and round holes are punched.

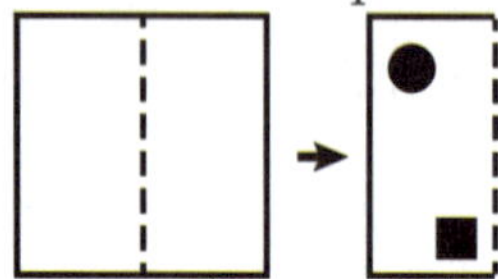

What does the sheet look like after it is unfolded again?

(A) 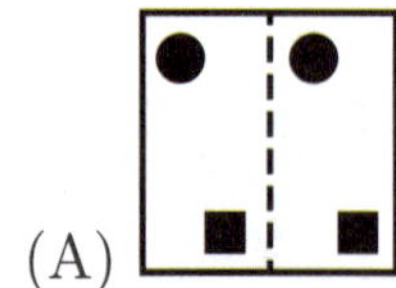 (B)  (C) 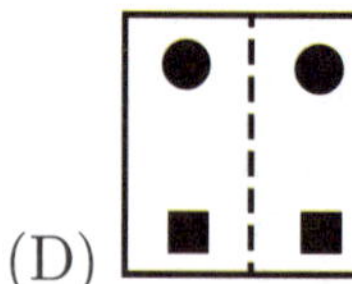 (D)  (E)

15. A student made the shape shown using 12 cubes. He put one drop of glue between any two cubes that share a common face. How many drops of glue did he use?

(A) 8 (B) 9 (C) 10 (D) 11 (E) 12

16. Max wants to complete the puzzle shown.

He has 5 different pieces, as shown. 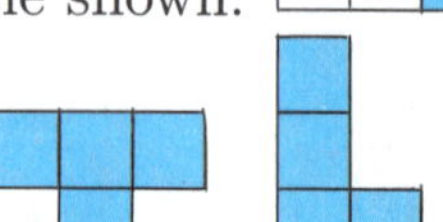 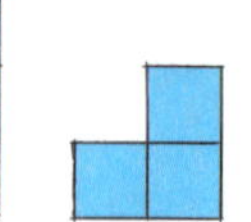
Which pieces does he have to use to complete the puzzle?

(A) (B) (C)

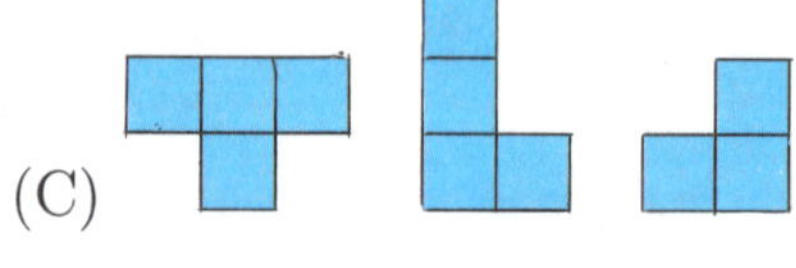

(D) (E)

Problems 5 points each

17. Arjun has 6 identical triangles like this:  . Which of the following pictures can he make?

(A) 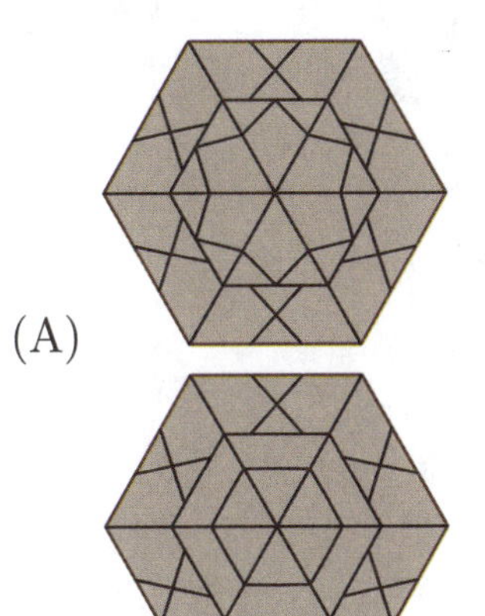 (B) 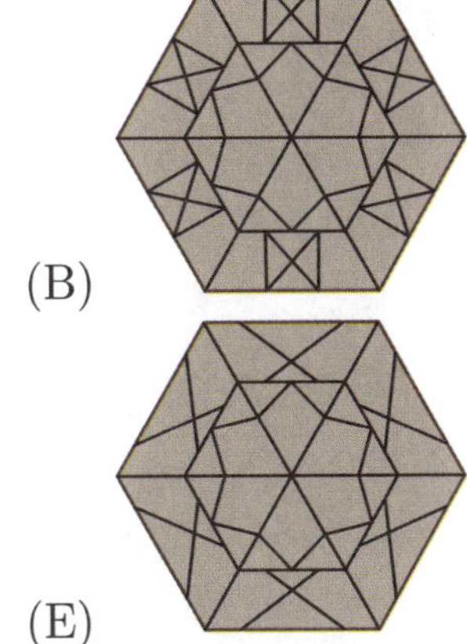 (C) 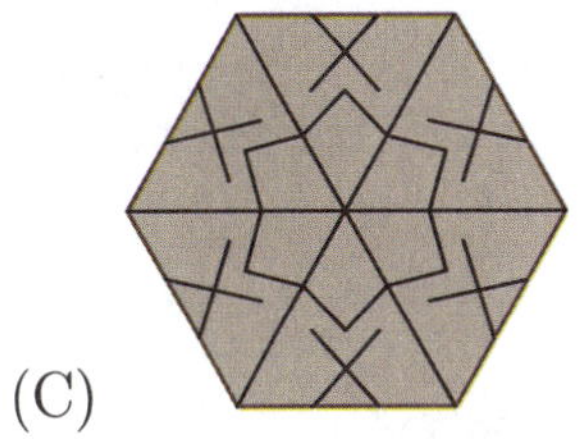

(D) (E)

18. Five children share a birthday and each child has their own cake. Lea is two years older than Jose but one year younger than Ali. Vittorio is the youngest. Which is Sarah's cake?

(A) 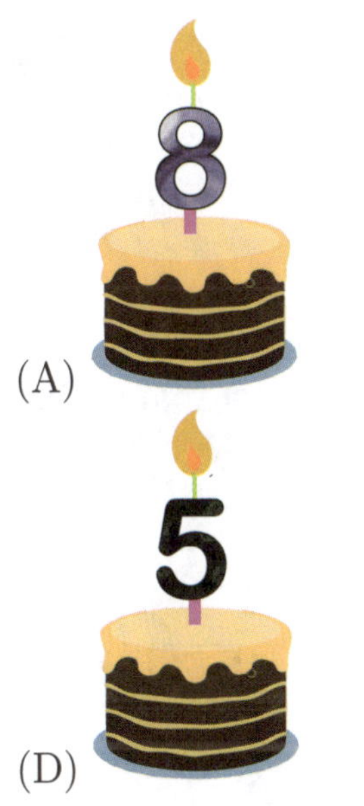

(B) 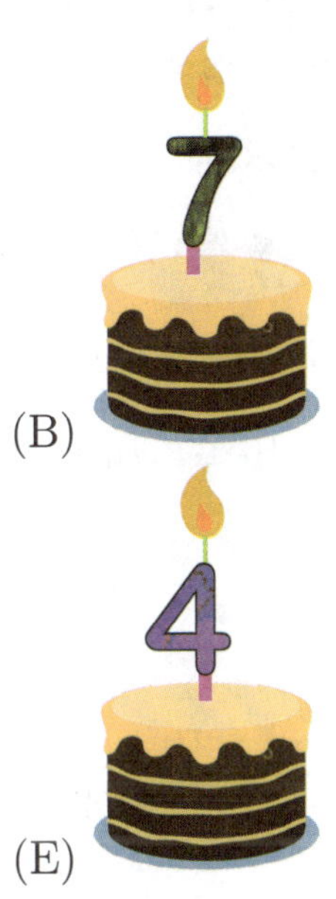

(C) 

(D) (E)

19. The map shows five villages A, B, C, D, and E, and the distances in kilometers between them. Only two villages are the same distance apart no matter which route you choose. Which are these two villages?

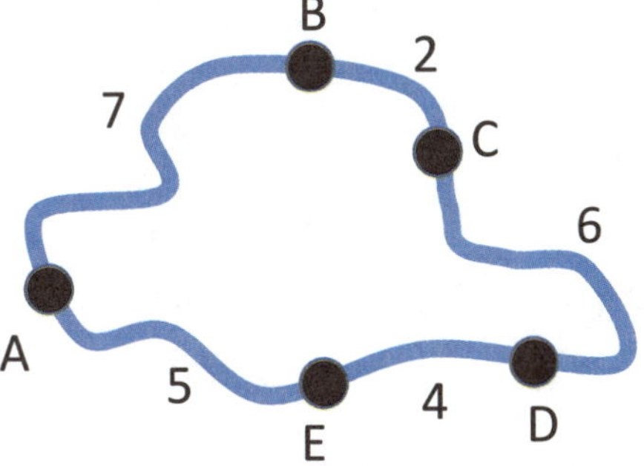

(A) B and E
(B) B and D
(C) C and E
(D) A and C
(E) B and E

20. Sam walks through a two-story maze from the entrance to the exit, both located at floor 1. In what order will she find the wall stickers?

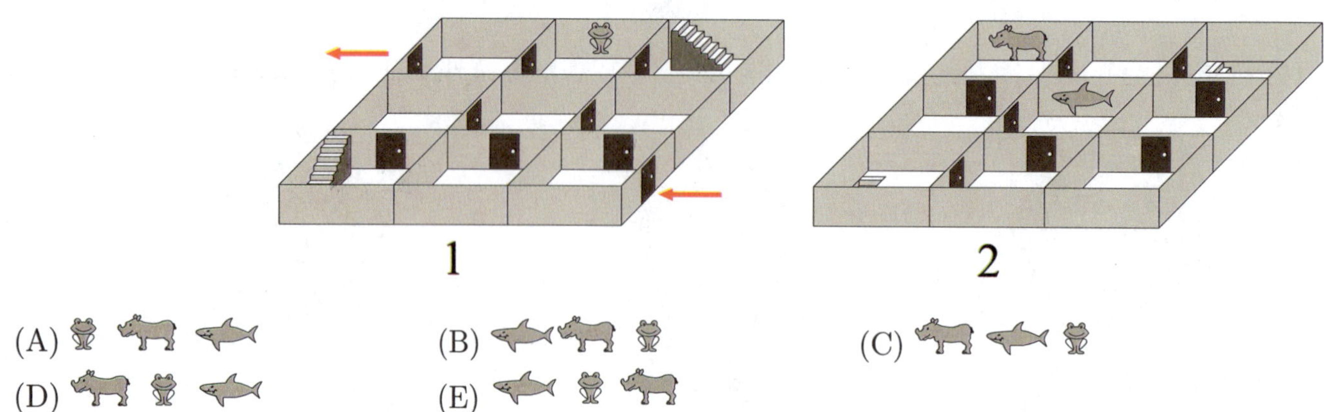

21. Emma finished third in a solo dance competition. There were three dancers between her and last place. In total, how many dancers took part in the competition?

(A) 4 (B) 5 (C) 6 (D) 7 (E) 8

22. Malik places one of the five pieces on the grid. He cannot rotate or flip the pieces. Which piece should he use to cover the numbers with the largest sum?

1	6	7
9	5	4
2	8	3

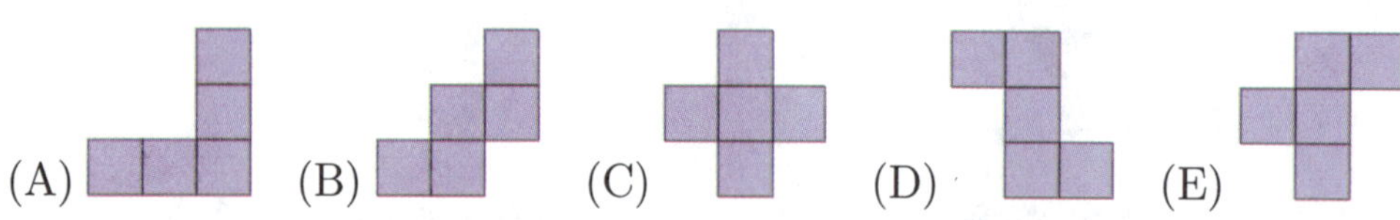

23. Three frogs live in a pond. Each night, one of the frogs sings a song to the other two. After 9 nights, one of the frogs had sung 2 times. Another frog had listened to 5 songs. How many songs had the third frog listened to?

(A) 7 (B) 6 (C) 5 (D) 4 (E) 3

24. Digits 1, 1, 2, and 3 are printed on four different cards. Three cards are laid out to make a subtraction, as shown in the picture. How many different results can we get?

(A) 6 (B) 8 (C) 10 (D) 12 (E) 24

Part II

Solutions

Solutions for Year 2005

1. (A) 4

Since 5 apples become ripe every day on each tree, we need to divide those 20 apples picked by the Green Gardener by 5 to get the number of the enchanted trees. $20 \div 5 = 4$

or

If there is one apple tree, there are 5 golden apples. If there are two apple trees, there are $5 + 5 = 10$ golden apples. If there are three apple trees, there are $5 + 5 + 5 = 15$ golden apples. If there are four apple trees, there are $5 + 5 + 5 + 5 = 20$ golden apples, which is the case in this problem.

2. (C) 4

Each person ate one of their two apples, so each of the four had $2 - 1 = 1$ apple left. Now, we have 4 people with one apple each, so they have 4 apples altogether.

3. (E) 7

4. (D) 6

The lips make an 8, the nose a 7, the eyes can be a 6 and a 9, the face is a 0, and there is a 2 on the hairline.

5. (C) 0

$5 + 2 = 7$
$7 - 3 = 4$
$4 - 4 = 0$

6. (C) 3

$5 - 4 + 3 - 2 + 1 =$
$1 + 3 - 2 + 1 =$
$4 - 2 + 1 =$
$2 + 1 = 3$

Or, we can group the quantities we are adding together.
$(5 - 4) + (3 - 2) + 1 = 3$
$1 + 1 + 1 = 3$

7. (C) 7

Michael is 4 years older than Ann. Since she is 3 years old, Michael is $4 + 3 = 7$ years old.

8. (D) 14

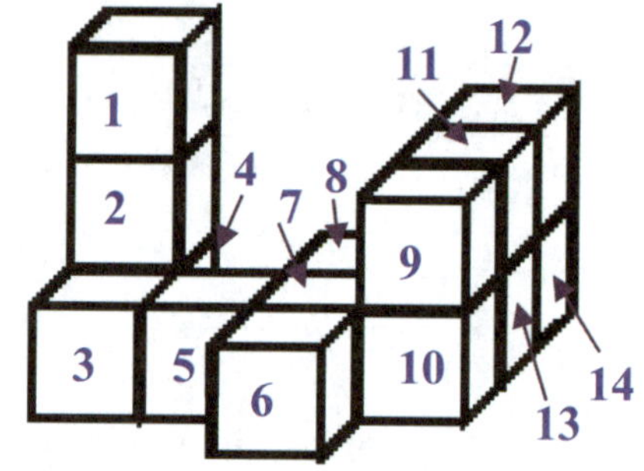

Count all the blocks individually as shown in the picture, or count the blocks going left to right row by row on the first level $(2 + 1 + 3 + 3 = 9)$, on the second level $(1 + 3 = 4)$ and on the third level (1). The total number of blocks is $9 + 4 + 1 = 14$.

9. (C)

Notice that going from left to right, the number of white beads to the right of each black bead increases by 1. The next set of beads will have 4 white beads. One bead is already shown on the string, so there will be three more white beads and then a black bead.

10. (C) 9

11. (A) 9

As she moves up three times, she will climb $2 + 2 + 2 = 6$ steps. Because she is already standing on the third step, we need to add 3 steps + 6 steps = 9 steps.

12. (A) 1

Every sheet has two sides, the first side with an odd number, and the other with an even number.

When we open a book to page 12, we should see

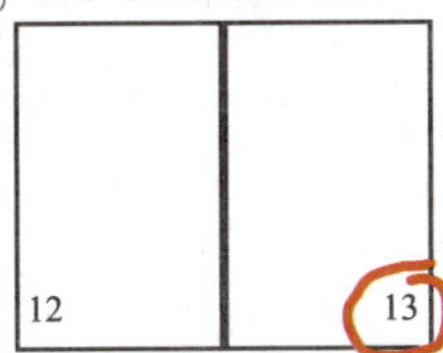

On the back of the page with the number 13 is page 14. Flipping that sheet, we should see

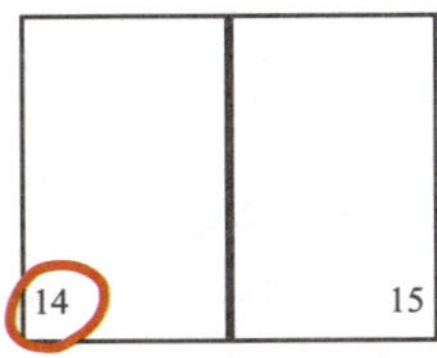

In the book in the problem, there are two numbers missing between 12 and 15, and these numbers are 13 and 14. So, only one sheet (with page numbers 13 and 14) is missing.

13. (B) 3

Two hens together will lay 2 eggs a day. In 2 days they will lay 4 eggs and in 3 days they will lay 6 eggs.

14. (C) 14

Horse: 2 horses × 4 legs each = 8 legs
Duck: 2 legs
Fish: 0 legs
Eagle: 2 legs
Boy: 2 legs
$8 + 2 + 0 + 2 + 2 = 14$

15. (A) 3

The two apples more that Maria has.

After taking away the two apples that are Maria's two additional apples, we can split the remaining 6 apples into 2 equal groups, which means Anne has 3 apples.

16. (A)

In every figure the number of rows increases by 1 and the number of columns increases by 2. Since the last figure has 3 rows and 5 columns, the next one has to have $3 + 1 = 4$ rows and $5 + 2 = 7$ columns.

17. (D) 9

Monika is 2 years younger than Karl $(4 - 2 = 2)$. When Karl is 11, Monika will be $11 - 2 = 9$ years old.

18. (B) 2nd

person who passed the winner of 3rd place = winner of 2nd place

19. (B) 6 kg

We know that the weight of the empty basket balances the 1 kg weight. So, the fruit will balance the remaining two weights, which together are 4 kg + 2 kg = 6 kg.

20. (C) +, +, –

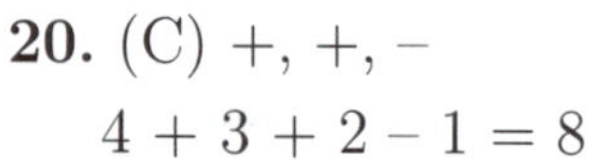

$4 + 3 + 2 - 1 = 8$

21. (E) 18

There are 7 digits (3, 4, 5, 6, 7, 8, 9) which added to 8 give more than 10.
There are 6 digits (4, 5, 6, 7, 8, 9) which added to 7 give more than 10.
There are 4 digits (6, 7, 8, 9) which added to 5 give more than 10.
There is 1 digit (9) which added to 2 gives more than 10.
There are no digits which added to 1 give more than 10.
Altogether, there are $7 + 6 + 4 + 1 + 0 = 18$ pairs of digits.

22. (A) 1

Notice that the number in the middle of each circle represents the difference between the number of figures in the shaded region and the number of figures in non-shaded region. The question mark in the last picture represents $4 - 3 = 1$

23. (D) 6

The diagonal from the lower left to upper right already has two of the three numbers. Since the sum must be 15, the missing number is $15 - 5 - 4 = 6$.

8	1	6
3	5	7
4	9	2

24. (E) white, red, green, yellow

We know the first car is white. The yellow car and the red car are not next to each other, so the green car has to be between them. So, the third car is green. The second car is not yellow, since the yellow car cannot be next to the white car. So, the second car is red and the last car is yellow.

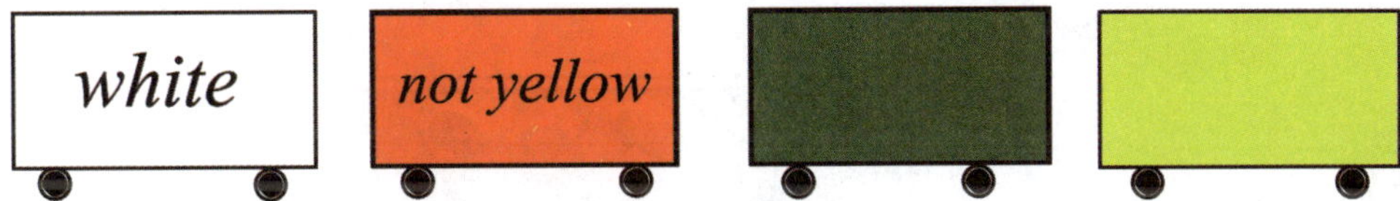

Solutions for Year 2007

1. (D) 

2. (B) 2 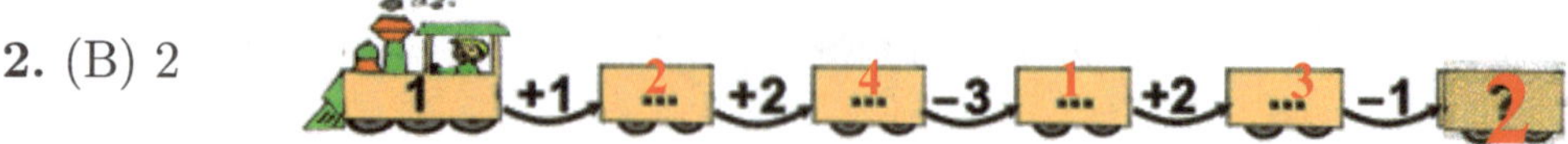

3. (C) 5

Each bicycle has one set of handlebars. I can see 5 handlebars in the picture, so there are 5 bicycles.

4. (A) **CAR**

$5 = \text{R}, 4 = \text{A}, 2 = \text{C};$

$2 < 4 < 5$

C < A < R makes the word "CAR."

5. (D) 

6. (D) 94

$101 - 7 = 101 - 1 - 6 = 100 - 6 = 94$

7. (C)

8. (A) 9 km

The distance from Behaiesti to the road sign is 4 km, and the distance from the road sign to Mormaiesti is 5 km, so the total distance between the villages is 4 km + 5 km = 9 km.

9. (C) O

SCHOOL, BOOK, PROBLEM, QUESTION

10. (D) Cristi is 7 years old.

Maria is 8 years old and Cristi is 7 years old, so only (D) is true.

11. (B) 20

You used 20 liters (50 liters – 30 liters = 20 liters).

12. (D) 14

Each cube has 6 square sides, so 3 cubes have 18 square sides altogether. Michael glued 4 squares to each other. He painted only the squares that were not glued, so he painted 14 squares (18 – 4 = 14).

13. (D) 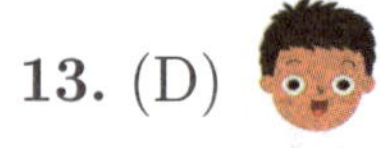

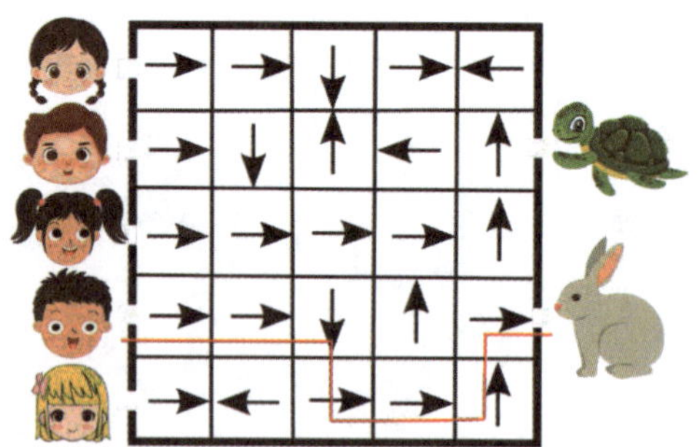

14. (C) 4

$$\begin{array}{r} {\scriptstyle 1} \\ 7 \\ 4 \\ +\,43 \\ \hline 54 \end{array}$$

$7 + 4 + 3 = 14$, $1 + 4 = 5$. The sum of the ones digits will be at least 10 ($7 + 3 = 10$) and no more than 19 (the largest digit is 9), you we will have to carry a one. The apple in the tens place in the third number will have to be 4 ($5 - 1 = 4$), so the apple represent the digit 4.

15. (A) 7

3 + 4 = 7

16. (A) \$15

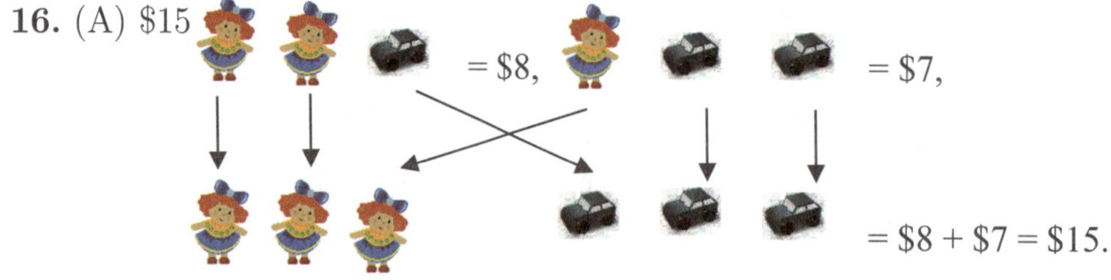

17. (D) 21

3 + 7 + 2 + 8 + 1

10 + 10 + 1

20 + 1 = 21

18. (C) 5 hours

Dan left school at 12:30 p.m. and walked home in 30 minutes, so he came back home at 1:00 p.m. It is 4 hours from 8:00 a.m. to 12 noon (12:00 p.m.), and 1 hour from 12:00 p.m. to 1:00 p.m. So, he was gone from his house for 5 hours $(4 + 1 = 5)$.

19. (B) 530

The digits are 0, 1, 2, 3, 4, 5, 6, 7, 8, and 9. The table below shows how to find all three-digit numbers such that the first digit is 2 more than the second digit, and the second digit is 3 more than the third digit. The table starts with the third digit, because it's the smallest.

Third digit	Second digit = Third digit + 3	First digit = Second digit + 2	Number
0	3	5	530
1	4	6	641
2	5	7	752
3	6	8	862
4	7	9	974
5	8	10	STOP! 10 is not a digit

The possible three-digit numbers are 530, 641, 752, 862, and 974. Only 530 is listed among the options, so (B) is the answer.

20. (E) Victor

21. (D) 24

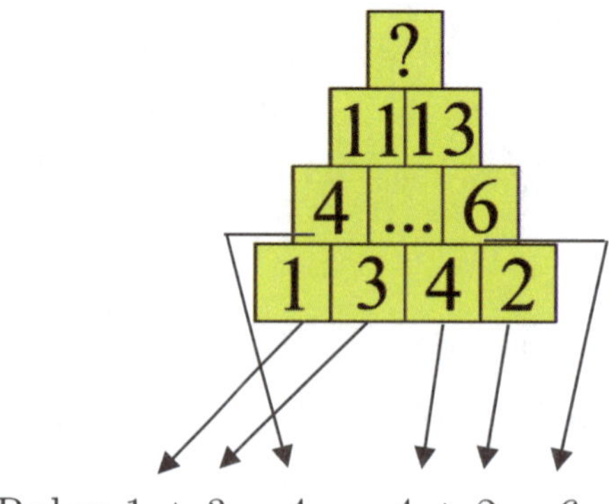

Rules: $1 + 3 = 4$, $4 + 2 = 6$,

so $11 + 13 = 24$.

If you add numbers from two neighboring boxes on one level, you get the number in the box above bordering the two boxes. For example, $1 + 3 = 4$, $3 + 4 = 7$ (this is the number that should be written where we see "..."), $4 + 2 = 6$, $4 + 7 = 11$, $7 + 6 = 13$, and $11 + 13 = 24$ at the top of the pyramid.

22. (E) 920

Notice that numbers 209, 902 and 290 are 3-digit numbers made using the digits 0, 2 and 9. The only other such number among the choices is 920.

23. (E) 8

Dan has 5 kinds of fruit-flavored candy (yellow, orange, red, green, and brown). The bag contained 20 pieces of candy, so there were 4 of each color ($4 + 4 + 4 + 4 + 4 = 20$). Dan ate 3 kinds of candy (yellow, red, and orange), so he had 2 kinds of candy left (green and brown)—exactly 8 pieces of candy ($4 + 4 = 8$).

24. (E) 9:30 to 4:30

If Andrea leaves at 4 p.m., she gets to the shopping center at 5 p.m., and half an hour after shopping center closes, so the shopping center closes at 4:30 p.m. If she leaves at 8 a.m., she gets to the shopping center at 9 a.m., and half an hour before the shopping center opens, so the shopping center opens at 9:30 a.m.

Solutions for Year 2009

1. (D) 10

There are 9 cubes in the base (3 rows and 3 columns) and one cube at the top. $(3 \times 3) + 1 = 9 + 1 = 10$.

2. (B) 11 $2 + 0 + 0 + 9 = 2 + 9 = 11$

3. (C) Ela

Count the number of candles on each birthday cake. Each candle represents one year of age.

7 years old

8 years old

9 years old

8 years old

4. (D)

(A)  (B)  (C) (D)

(A)	(B)	(C)	(D)
3 apples 2 pears	3 apples 3 pears	5 apples 3 pears	3 apples 4 pears

5. (B) 8

Since all the numbers add up to 50, add the numbers you can see, and then subtract from 50 to get the hidden number. $5 + 20 + 17 = 42$. $50 - 42 = 8$.

6. (A) 28 

Paul

12 cars

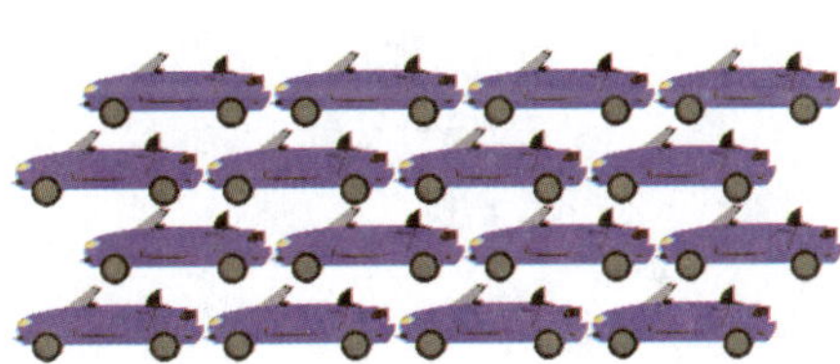

$12 + 4 = 16$ cars

$\mathbf{12 + 16 = 28}$

7. (C) 13

1 apple for Snow White + 1 apple for the Prince + 7 apples for the 7 Dwarfs + 4 apples in the basket = 13 apples

8. (A) 3

Ones column	Tens column
− 1 = 2	4 − = 1
= 3	= 3

9. (C) $39

$12 + 3 × $9 = $12 + $27 = $39

One adult ticket Three children's tickets

10. (D) 27

First, find what number covers. 21 − 7 = 14. Then, substitute 14 into the second equation.

$2 \times 14 = \square + 1$

$28 = \square + 1$

$28 - 1 = \square$

$27 = \square$

11. (D) Thursday

Ada will be taking the pills for 60 days. There are 8 full weeks in 60 days, with 4 days left over (60 days = 8 × 7 days + 4 days). If Ada starts with Monday, the fourth day of the week is Thursday.

days 1-56 (8 weeks)

Monday	Tuesday	Wednesday	Thursday	Friday	Saturday	Sunday
day 57	**day 58**	**day 59**	**day 60**			

12. (B) 54

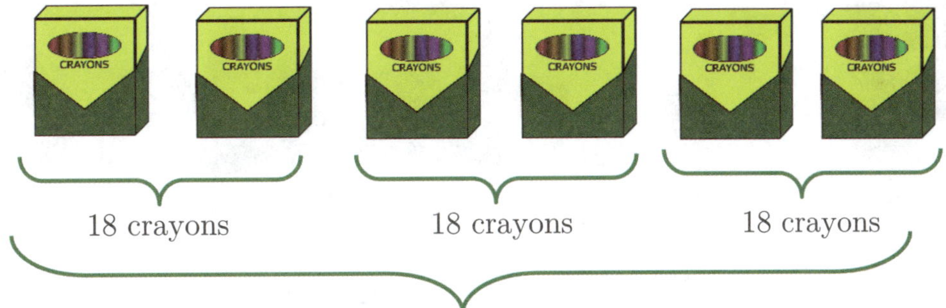

18 crayons + 18 crayons + 18 crayons = 54 crayons

13. (B) 3 inches

14. (C)

Ilona drew $6 \div 3 = 2$ flowers.
She drew $4 + 2 = 6$ hearts.

15. (C) 7

The greatest number is 8. The smallest number is 1.
The difference between them is $8 - 1 = 7$.

16. (C)

At 6:00 (six o'clock), the hour hand (the shorter hand) needs to point to 6, and the minute hand (the longer hand) needs to point to 12.

17. (B) 4

With 4 chimpanzees and 3 baboons, we need 12 gorillas to have 19 animals.
Since the 12 gorillas were in 3 cages, there were 4 gorillas in each cage ($3 \times 4 = 12$).

18. (D) 34

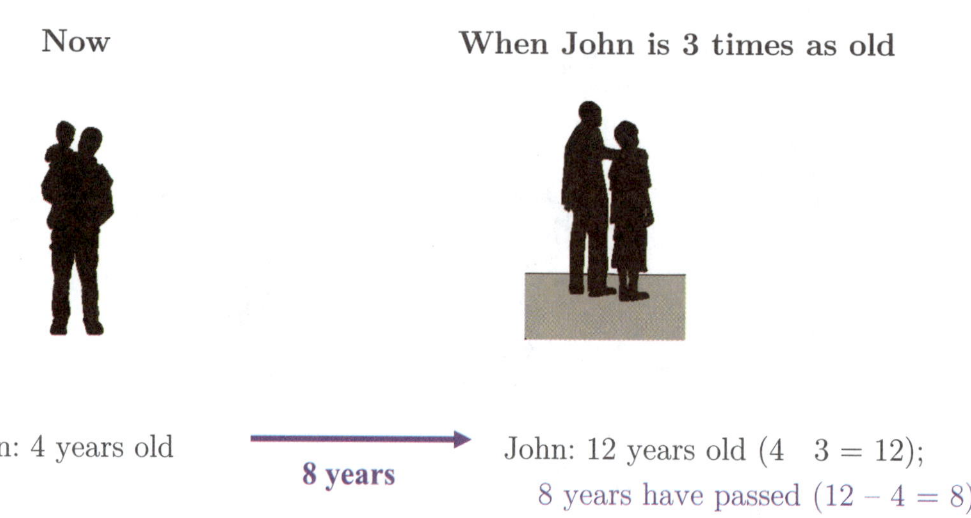

John: 4 years old → John: 12 years old (4   3 = 12); 8 years have passed ($12 - 4 = 8$)

Father: 26 years old → Father: 34 years old ($26 + 8 = 34$)

19. (A) 10

Notice that there were 11 more blueberry dumplings than cheese dumplings.
If the number of blueberry dumplings and the number of cheese dumplings were equal, there would be 11 more cheese dumplings and the total would be $31 + 11 = 42$ dumplings.
The number of blueberry dumplings is half of 42, so there are 21 blueberry dumplings.
$31 - 21 = 10$ is the number of cheese dumplings.

20. (D) \$3

The information that Eva already bought two notebooks is not necessary for solving this problem.

two notebooks cost:  \$4 (how much Eva has) + \$2 (how much Eva needs) = \$6

one notebook costs:  $\$6 \div 2 = \$3$

21. (C) Paul

Least stamps ⟶ **Most stamps**

Paul Tom Matt? Adam?

The boy with the smallest number of stamps is *not* Matt, because Matt has more stamps than Paul. The boy with the smallest number of stamps is *not* Adam, because Adam has more stamps than Tom. The boy with the smallest number of stamps is *not* Tom, because the problem states this. So, Paul is the boy with the smallest number of stamps.

22. (B) 17

Mother cleaned:

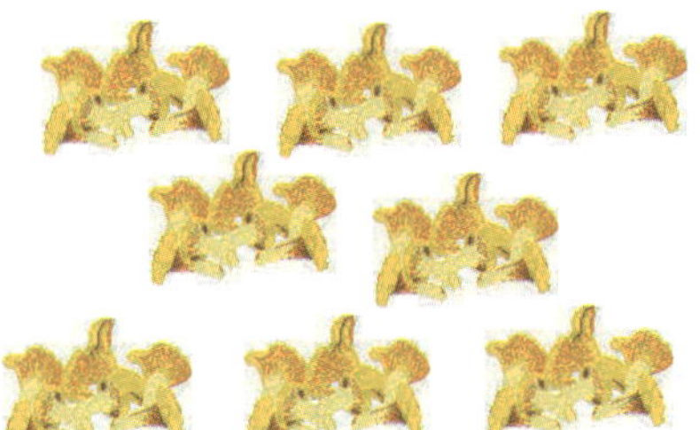

7 mushrooms in 5 minutes ⟶ **56 mushrooms in 40 minutes**

There are 8 5-minute intervals in 40 minutes (8×5 minutes = 40 minutes).

8×7 mushrooms = 56 mushrooms

Father found:

56 mushrooms total: 39 in the first hour and **56 − 39 = 17 during the second hour**.

23. (C) 5

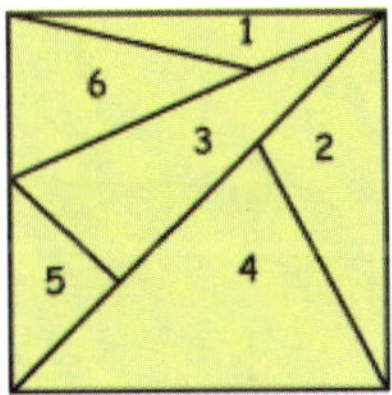

24. (B) 6

There are 6 different 3-digit numbers that we can make using the digits 1, 2, and 3:

123	**213**	**312**
132	**231**	**321**

Solutions for Year 2011

1. (C) 3
These are the numbers we use when we count in order. So, the boxes should have the numbers 1, 2, 3, 4, and 5. The missing number is 3.

2. (D) 8 $6 + 2 = 8$

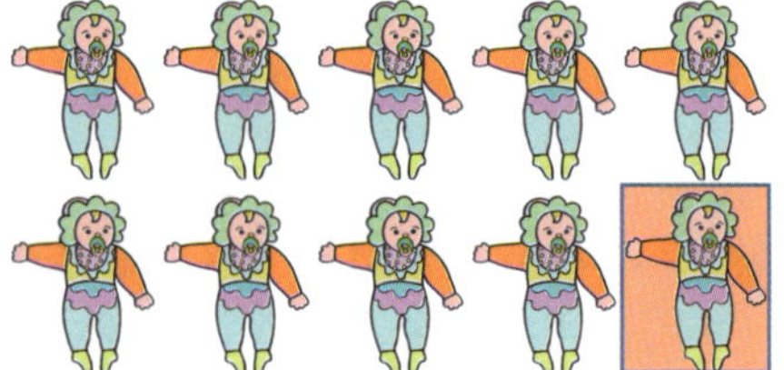

3. (D) 9
Take one away from ten, $10 - 1 = 9$. Sharon now has nine dolls.

4. (A) 12
Each boy has two legs, and each dog has four legs. The boys have a total of $2 \times 2 = 4$ legs, and the dogs have a total of $2 \times 4 = 8$ legs. Altogether, there are $4 + 8 = 12$ legs on the playground.

5. (B) February
February is the only month of the year which sometimes has exactly 29 days. This happens every four years when there is a leap year. In other years, it has 28 days.

6. (B) 1
7 students will each get one candy bar and one glass of milk. That leaves $8 - 7 = 1$ candy bar for the teacher to get with his coffee.

7. (C) 4
There are four digits in the number 2011. When we add them we get $2 + 0 + 1 + 1 = 4$.

8. (B)
There are 3 dolls wearing dresses, 2 dolls with two braids, but only 1 with a dress, two braids, and one flower.

9. (C) 24
There are two boots in one pair. 12 pairs of boots is $12 \times 2 = 24$ boots.

10. (A) 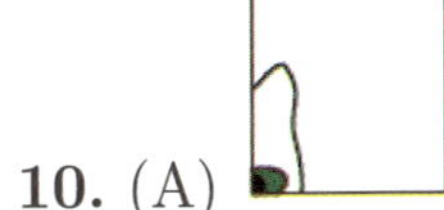

We need a piece of the puzzle which has a part of a cat's eye AND ear.

11. (D) 2/25/2011

Remember that we write dates with the month first, then the day, and then the year. Today is March 12, 2011. The expiration dates for the items are:

(A) September 15, 2011 (B) March 4, 2012 (C) July 11, 2011 (D) February 25, 2011

March 12, 2011 is after February 25, 2011, so (D) is the item that is past its expiration date.

12. (B) 64

Mark's grandmother will be 100 years old 36 years from now. This means that now she is $100 - 36 = 64$ years old.

13. (B) 2

A cat has 2 ears and a dog has 4 paws. Anne has 4 cats, so they have $4 \times 2 = 8$ ears. Each dog has 4 paws, so there are 2 dogs if there are 8 paws, because $2 \times 4 = 8$.

14. (A) 

Start with top triangle and follow the path as shown:

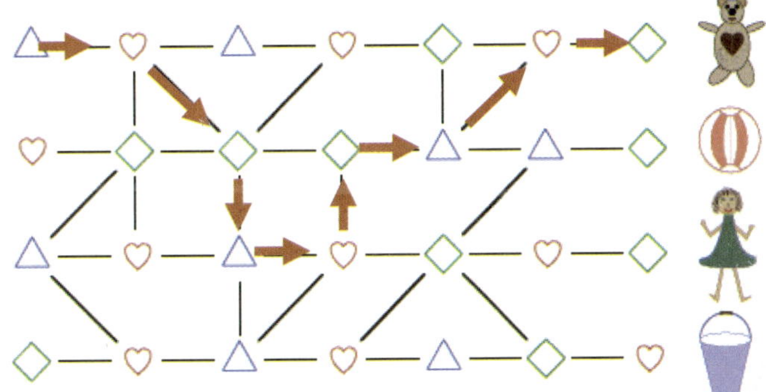

15. (B) 12:00

The next train listed leaves at 9:15. The trip takes 2 hours and 45 minutes, so he will arrive in Indianapolis at 12:00. (The trip starts 45 minutes before 10:00, and then it's two full hours until 12:00.)

16. (C) \$1.60

Both girls bought two pens and two erasers. The difference between what Katie and Hannah bought is that Katie bought 3 pencils and Hannah bought only 1. That is a difference of 2 pencils. The difference in price is \$11.60 – \$8.40 = \$3.20. So, two pencils cost \$3.20, and one pencil costs \$1.60.

17. (A)

After the piece of paper is unfolded it will have an "unbroken" figure in the center.

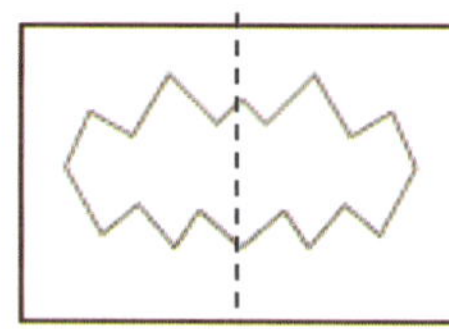

18. (D) 15

The youngest daughter is 5 years old. The middle daughter is 6 years older than the youngest, so she is $5 + 6 = 11$ years old. If the middle daughter is 4 years younger than the oldest daughter, the oldest daughter is 4 years older than the middle daughter. So, the oldest daughter is $11 + 4 = 15$ years old.

19. (D) 4

There were $16 + 11 + 17 = 44$ flowers at the beginning. The owner sold bouquets of 5 flowers, so she sold a number of flowers that is a multiple of 5. The largest multiple of 5 less than 44 is 40, so there were $44 - 40 = 4$ flowers left.

20. (D) 8

To make the amount of water the same in both aquariums, Simon has to pour half of the difference into the first aquarium. First we notice that there are $42 - 26 = 16$ quarts of water more in the second aquarium. Half of the difference is 8 quarts.
Check that $26 + 8 = 34$ and $42 - 8 = 34$.

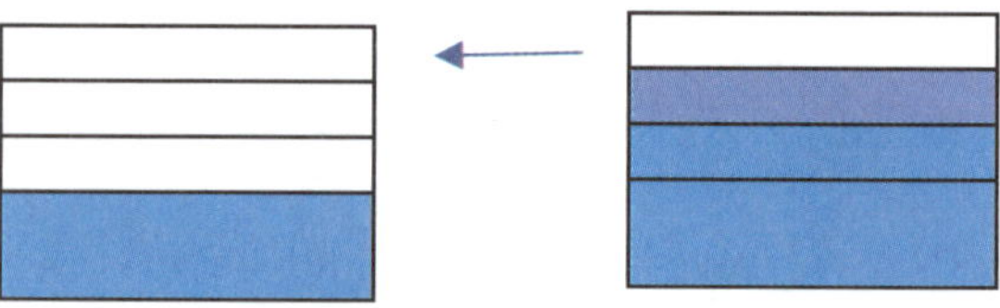

You need to move half of the difference from one aquarium to the other to make the amount of water equal.

21. (B) 4 lbs

We can start with:

 +  = 11 lb

 +  = 11 lb – 1 lb = 10 lb

(continued on next page)

Put these two statements together, adding all the animals on one side and getting the total weight on the other:

Compare with this statement given in the problem:

The only difference is one monkey, so we see that one monkey weighs 24 – 21 = 3 lb. Use this information in the second statement. If Philemon the Cat and two monkeys weigh 10 lb, then Philemon weighs 10 lb – 3 lb – 3 lb = 4 lb.

22. (A) Anita

Daniel ate more apples than Clara, so Clara did not win. Michael ate fewer apples than Anita, so Michael did not win. We already know that Daniel did not win. So, Anita ate the most apples and won.

23. (D) 12

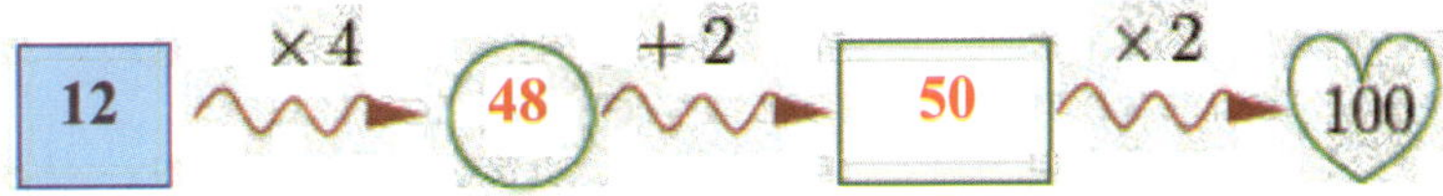

Work backwards.

Write 50 in the rectangle because 50 × 2 =100.

Write 48 in the circle because 48 + 2 = 50.

Write 12 in the square because 12 × 4 = 48, so 12 is the answer.

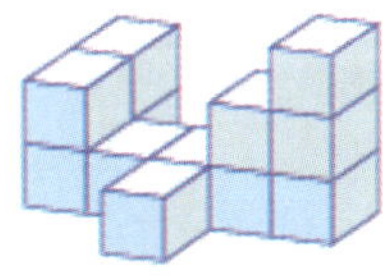

24. (B)

Make sure that the number of blocks placed one on top of another is as the diagram shows. Answer (A) has 4 blocks one on top of another and answer (D) has only 2 blocks at most one on top of another, but there are 3 blocks one on top of another on the far right of the figure. Also, Jon's figure measures 5 blocks from left to right which eliminates answer (C). (B) fits the description of his building plan.

Solutions for Year 2013

1. (D) 6 and 9

You can find the digits: 0, 1, 2, 3, 4, 5, 7, and 8. The missing digits are 6 and 9.

2. (B) 8

Each of 4 children takes one book, which means 4 books are taken from 12 total on the shelf. There will be 8 books left on the shelf since $12 - 4 = 8$.

3. (A) 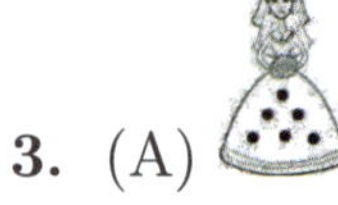

6 is the only whole number that is less than 7 but more than 5. The dress with 6 dots is marked with letter (A)

4. (D) 

Picture (D) has 3 black kangaroos and only 2 white kangaroos. 3 is greater than 2.

5. (B) 5

There are 5 more bricks in the larger stack as marked in black below:

6. (E) 9

9 tiles fit in the area inside as shown in the picture.

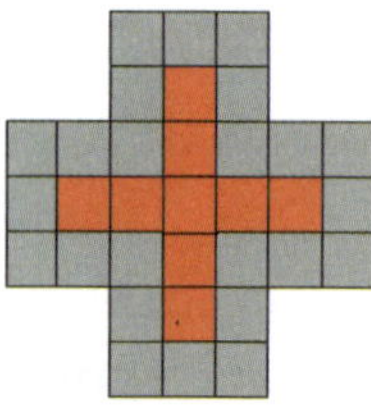

7. (C)

Note that the cut out piece must have 3 buttons, one star, a line of frosting, and cannot have a heart.

8. (D) 

Ann has earrings so (C) is not Barb. Eve has a necklace so (B) is not Barb. Jim has a hat so (E) is not Barb. Bob has glasses so (A) is not Barb. The only person left is Barb, marked with (D).

9. (E) 9

The pictures below show the way the children distribute the apples among themselves.

After Ana gives 3 apples to Sanja:

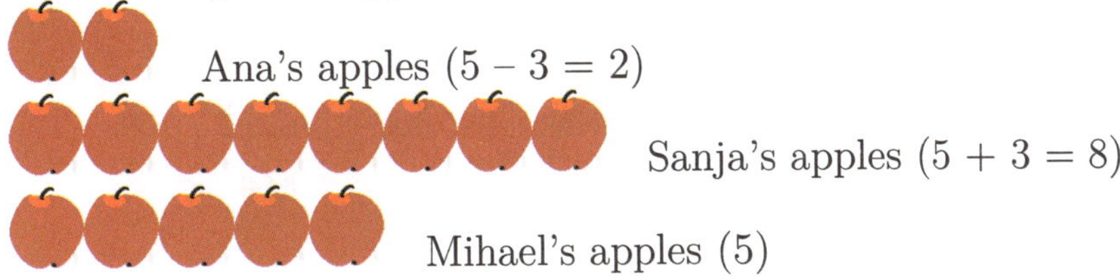

After Sanja gives half of her apples to Mihael:

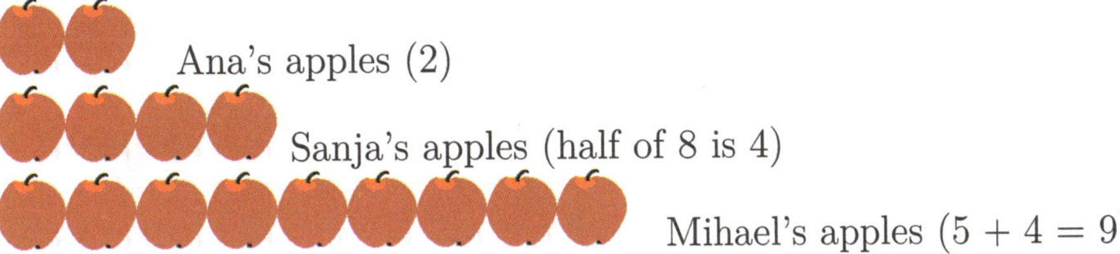

10. (C) 3 kilograms.

George and his 2 cats weigh 36 kg. Since George weighs 30 kg, then 2 cats together must weigh 6 kg ($36 - 30 = 6$). Since both cats are the same weight, then each cat must weigh 3 kg ($3 + 3 = 6$).

11. (A)

The pattern made by the first four squares is repeated 6 times. Then there is another gray square at the end. So, there are 7 gray squares. The other kinds of squares appear 6 times each.

12. (B) 8

The picture shows the carrots the rabbit can eat walking freely in this maze.

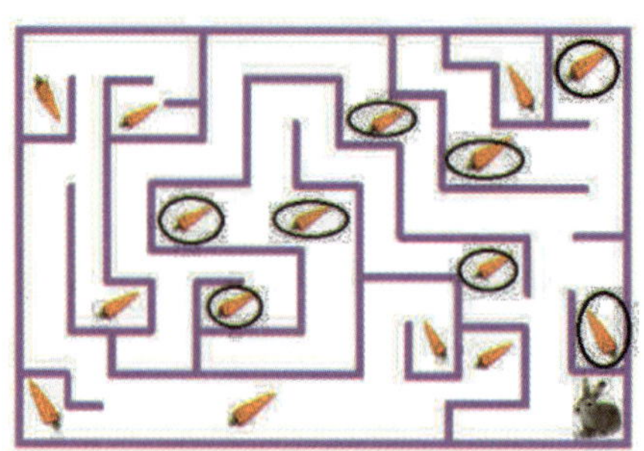

13. (D) 4

The picture below shows the Cat's and Mouse's jumps. After six jumps Cat catches Mouse on tile number 4.

14. (E) 24

Notice that the front layer has 12 cubes and the back layer has 12 cubes. The total number of cubes is $12 + 12 = 24$.

15. (C) Dannie

Kitty is 2 years older than Betty. Annie is the same age as Betty (they are twins) so Kitty is also 2 years older than Annie. Teddy is 3 years older than Annie, so he is 1 year older than Kitty. Since Dannie is 2 years older than Kitty, he is 1 year older than Teddy. Dannie is the eldest as shown in the picture below by lining up all five children from the oldest to the youngest.

16. (A)

Each letter A turns clockwise the same amount (half of a quarter turn). The next letter A would turn again and look like answer (A).

17. (E) 2 brothers and 4 sisters

Since Kasia has 3 brothers and 3 sisters then Mike has to have 2 brothers and 4 sisters as illustrated in the picture below:

Kasia (Mike's sister)	Brother 1 (Mike)	Brother 2	Brother 3	Sister 1	Sister 2	Sister 3

18. (C) 18

Since 1 pear is equal to 2 apples then 6 pears are equal to 12 apples.

Since one apple is equal to 3 plums, then 12 apples will equal 36 plums.

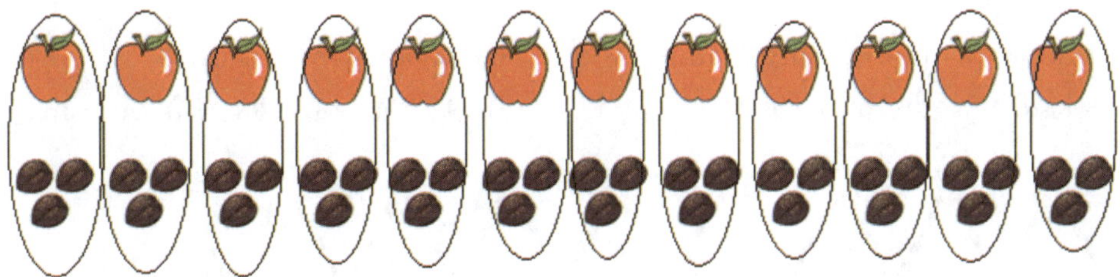

Since one strawberry is equal to 2 plums then 36 plums will equal 18 strawberries.

19. (C) 3

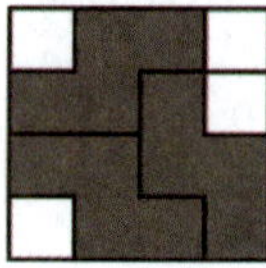

Ann can cut 3 pieces out of the sheet in a few different ways. One way is shown on the right.

It is not possible to cut four pieces from the sheet. Here is one way to show why not: Each piece is made up of 4 small squares. Four such pieces are made up of 16 small squares, which is exactly the area of the sheet. If we could make four pieces, we would use every small square on the sheet, including all four corners. Let's start by cutting a piece out at the upper left-hand corner. There are two ways to do it as shown below.

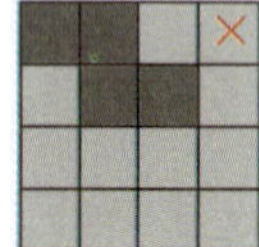

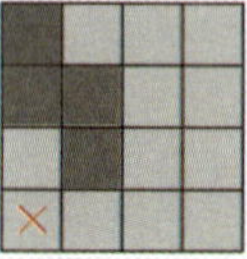

Notice that if we place the piece horizontally, we cannot cut another piece that would use the upper right-hand corner. If we place it vertically, we cannot cut another piece that would use the lower left-hand corner. Thus, it is not possible to cut out pieces and use all 16 small squares, and so it's not possible to cut out four pieces.

20. (B) 51

The first house is made using 6 matchsticks; each additional house uses 5 more matchsticks.6 matchsticks for the first house + 9 more houses × 5 matchsticks each = $6 + 45 = 51$ matchsticks.

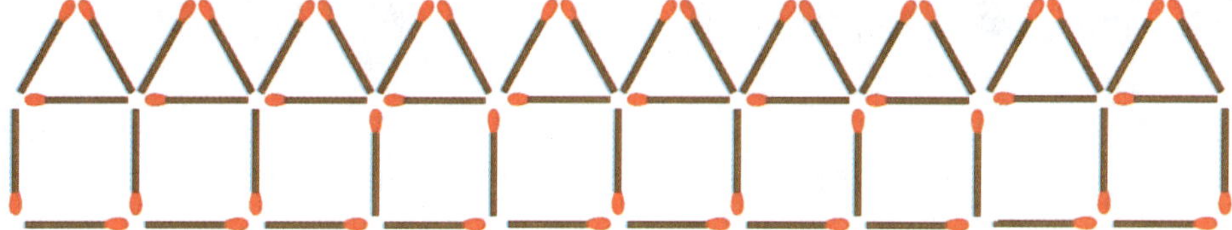

21. (A) only b

Only tile b matches the pattern as picture below shows (the tile has to be turned counterclockwise to fit the empty spot):

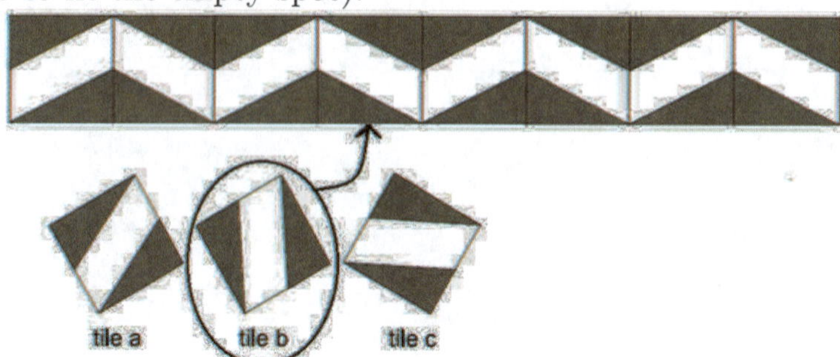

22. (D) 15

With coins: 5 cents, 10 cents, 20 cents, and 50 cents, Ana can make the 15 following different values: 5, 10, 15, 20, 25, 30, 35, 50, 55, 60, 65, 70, 75, 80, and 85.
Using only one coin, she can get 5, 10, 20, 50. Using two coins, she can get $5 + 10 = 15$; $5 + 20 = 25$; $5 + 50 = 55$; $10 + 20 = 30$; $10 + 50 = 60$; and $20 + 50 = 70$. Using three coins, she can get $5 + 10 + 20 = 35$, $5 + 10 + 50 = 65$, $5 + 20 + 50 = 75$, and $10 + 20 + 50 = 80$. Using all four coins she gets $5 + 10 + 20 + 50 = 85$.

23. (D) 4

From the picture we can see which four corner cubes were removed. For some of the stamps, the cube needs to be rotated.

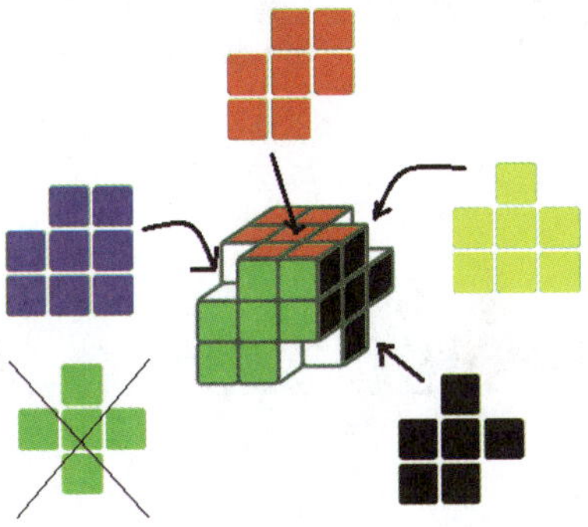

24. (D) 52

In order for 20 pieces to be along the walls of a square box, each layer has to contain 36 chocolate pieces (a 6 by 6 square) as illustrated in the picture. In the upper layer, after 20 chocolate pieces are eaten, only 16 remain ($36 - 20 = 16$). In the bottom layer, however, there are still 36 chocolate pieces. The total number of chocolate pieces left in the box is: $16 + 36 = 52$.

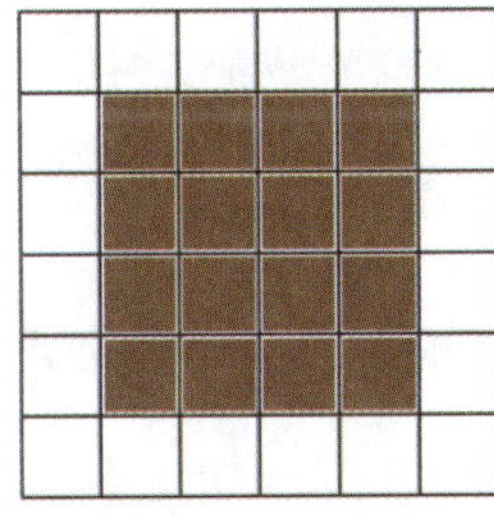

Upper layer

Lower layer

Solutions for Year 2015

1. (D) 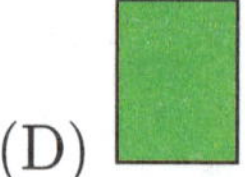

Let's draw the figures used in each picture separately and in the same order. We see that the second picture is missing the green rectangle.

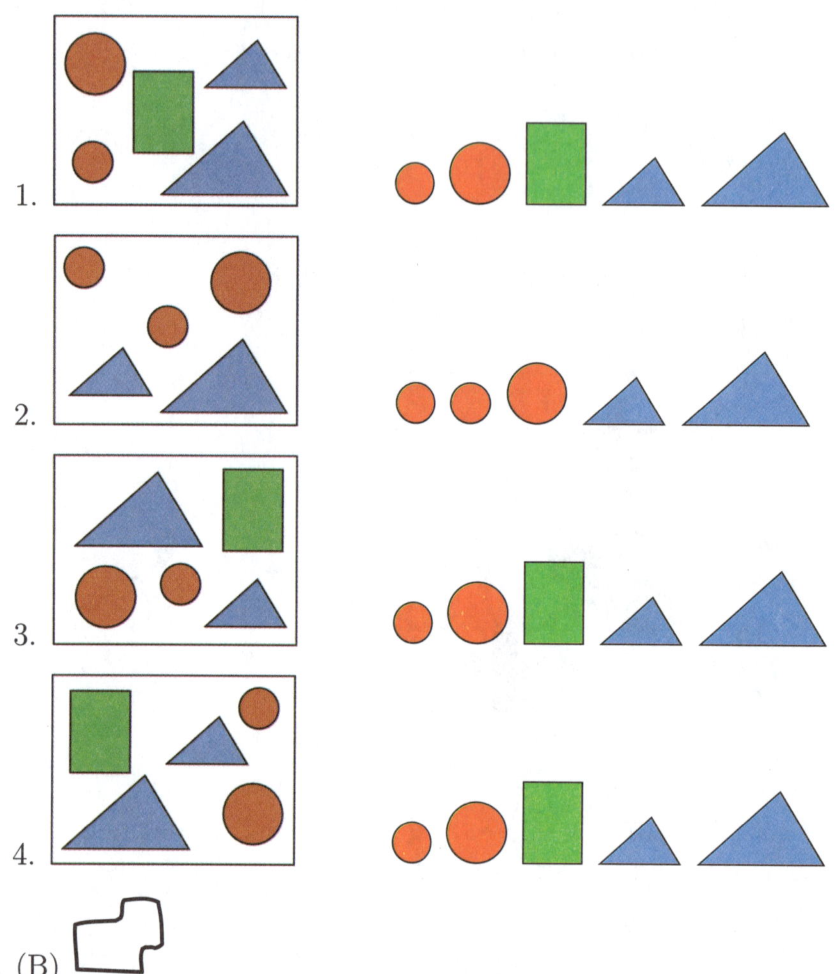

2. (B)

Trace the missing piece of the roof and color it red. The shape matches the shape in choice (B) if it's rotated.

3. (C) 19

Count the number of spots on each ladybug. The picture on the right shows how many spots are on each ladybug. Add all the numbers together:

$$2 + 3 + 3 + 5 + 6 = 19$$

4. (E) 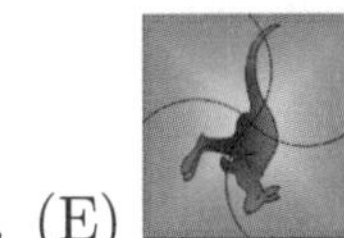

We rotate the original picture by 90° several times until we see all possible rotations.

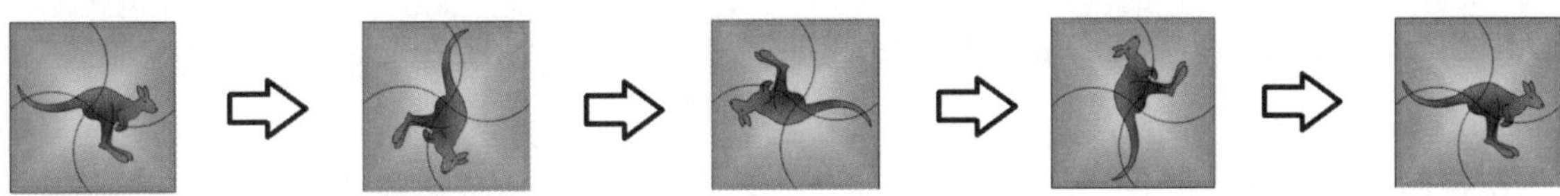

We see that only the picture in choice (E) appears as a possible rotation of the original picture.

5. (B) 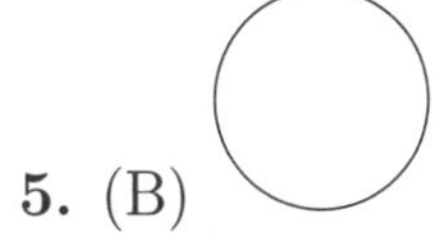

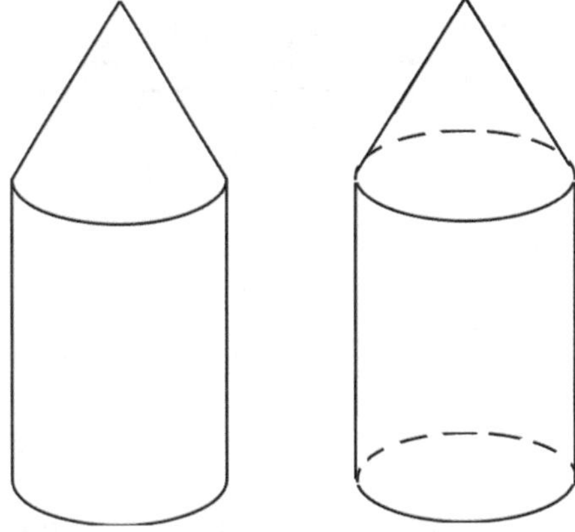

The bottom part of the tower looks like a cylinder and the top part looks like a cone. The cone has the same circular base as the top of the cylinder. The picture on the right shows the hidden back portion of the cylindrical piece in dashed lines. Looking at the tower from above, we can only see the circular shape of the cylinder.

6. (E) 10

The first column of numbers, 2 and 8, is inside the blue circle, but outside the red square. The remaining two columns are inside the red square. We add the numbers in the first column to get the answer:

$$2 + 8 = 10$$

7. (C) 1 hour

Half of the whole movie lasts half of a whole hour. The whole movie will last twice as long. Two halves make a whole, so the whole movies lasts the whole hour.

8. (B)
The shortest strip Eric makes will be the one with the largest overlap between the two connected strips. Count the number of holes between the two screws connecting the two metal strips in each strip. Strip B has the longest overlap, so it is the shortest.

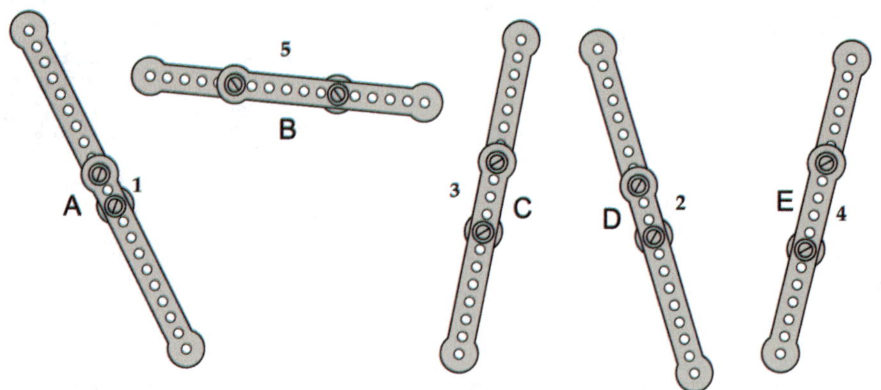

You can also count the total number of holes you see in each long strip.

9. (D) 40 meters
There are 10 spaces between the 11 flags. Each space is 4 meters long, so all together the track is as long as the combined distance between all the flags. We multiply 10×4 meters $= 40$ meters. The track is 40 meters long.

10. (C) 4
To make the numbers of pieces of candy equal, Tomo needs to give Marko half of the difference in the number of pieces of candy. The difference is $17 - 9 = 8$. Half of 8 is 4. Check that 4 is the answer: $9 + 4 = 13$ and $17 - 4 = 13$.

11. (C) 12
Since cubes of the same color do not touch, and each tower has a gray cube on top as we can see in the picture, then each of the towers is exactly the same as the picture on the right. We see there are 2 white cubes and 3 gray cubes in each tower. In the total of 6 identical towers there will be $6 \times 2 = 12$ white cubes.

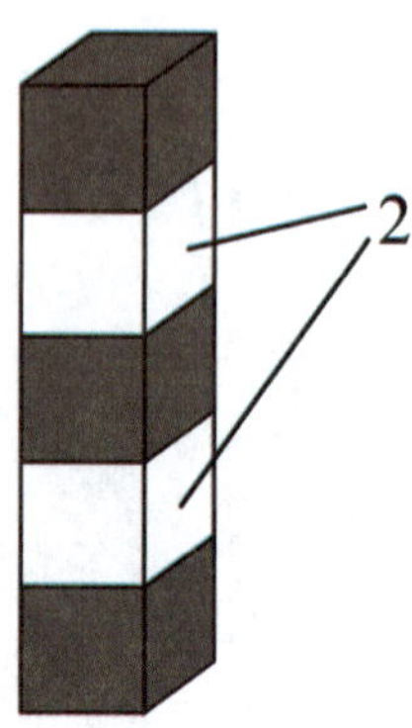

12. (E) May fifteenth, 2015

The dates we are given can be written as

1. May tenth, 2015, in (A) as 5/10/2015
2. April twenty-fifth, 2015, in (B) as 4/25/2015
3. May twenty-fifth, 2015, in (C) as 5/25/2015
4. January fifth, 2055, in (D) as 1/5/2055
5. May fifteenth, 2015, in (E) as 5/15/2015

The only dates that will have three 5's are the last three on the list. The 2015 dates will occur first, and May 15th is earlier than May 25th.

13. (E) 5

If 5 is to the left of the question mark, then the sum written in the box with the question mark is greater than 5, but only numbers from 1 to 5 are available. If 5 is to the right of the question mark, then read the subtraction in opposite direction, so it becomes addition. Again, the sum would be greater than 5. So, 5 must be placed in the box with the question mark. This can be done in many ways as shown below.

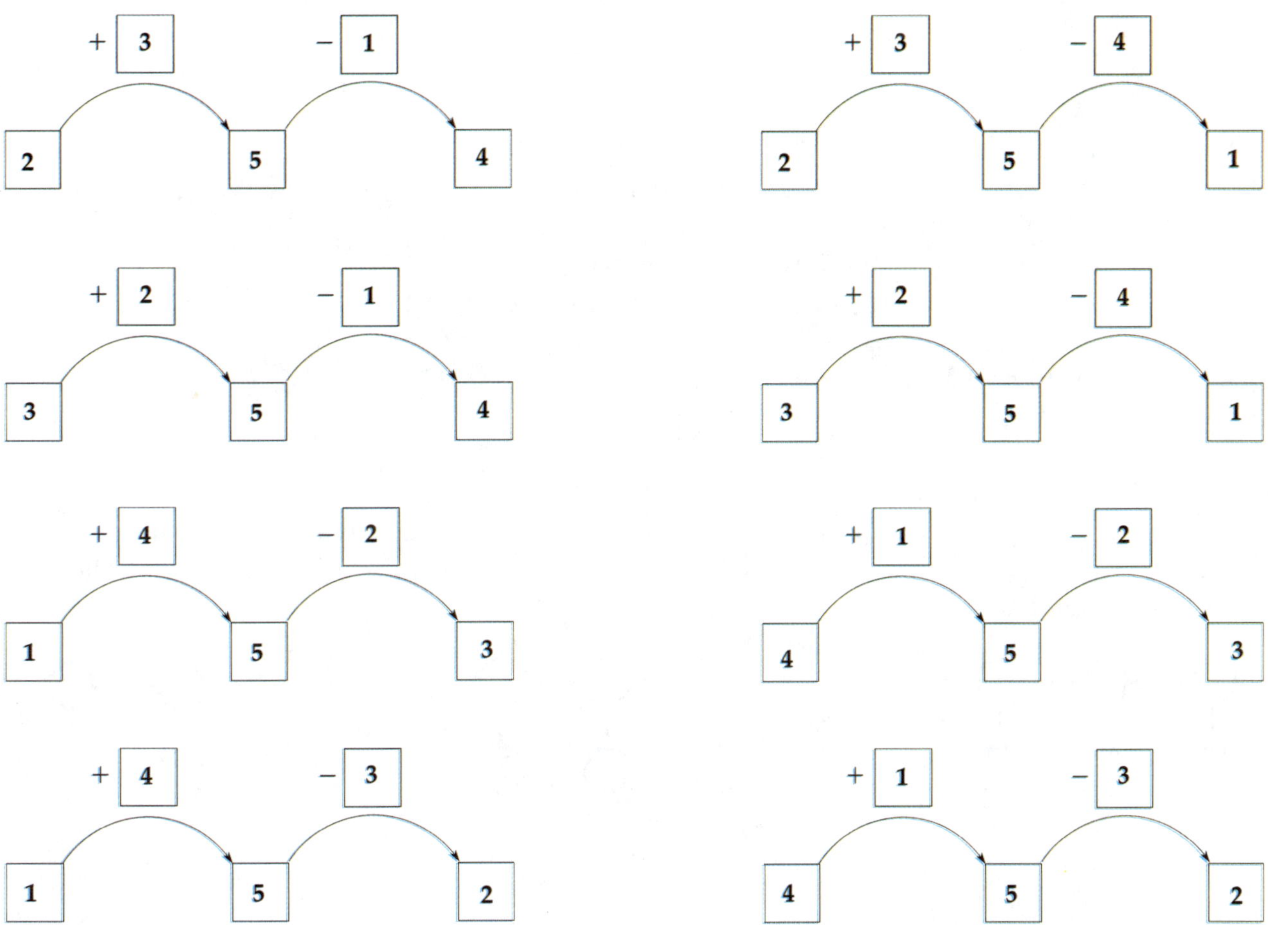

14. (D) 2

Each of the 2 pizzas was cut into 8 slices for a total of $8 + 8 = 16$ slices. Each of the 13 guests had a slice of pizza and so did Vera. From the 16 slices of pizza, 13 were eaten by the guests and 1 slice was eaten by Vera. Subtract to find out how many slices remain. There were $16 - 14 = 2$ slices of pizza left over.

15. (C)

We can split up Figure (C) into only one brick identical to the bricks Don has. In each case we are left with some bricks that are cubes. So Figure (C) cannot be built using the two bricks Don has.

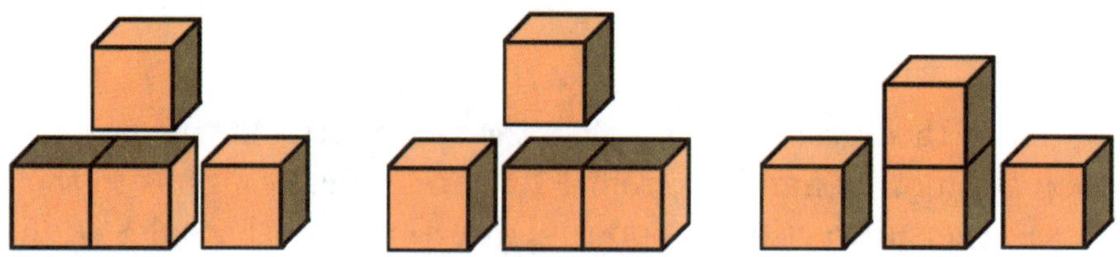

16. (A)

We can begin drawing in the missing piece starting with the star with a single outline. Counterclockwise to the star with the single outline is a star with a double outline, as shown in the picture on the right, and counterclockwise to it is a star with a triple outline. Only answers (A) and (B) have these three types of stars. Piece (B) has the opposite or clockwise arrangement of the stars. Piece (A) has them arranged counterclockwise.

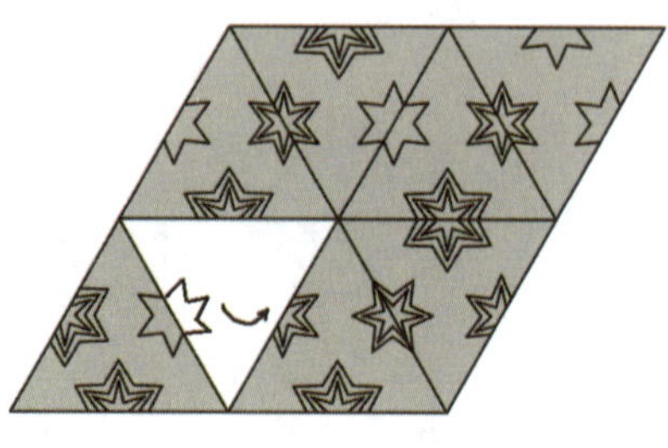

17. (D) 6

Considering just the upper portion of the diagram we have the three paths shown below. By symmetry, there are three more paths Jake the Kangaroo can take by jumping along the lower portion of the diagram, for a total of 6 paths.

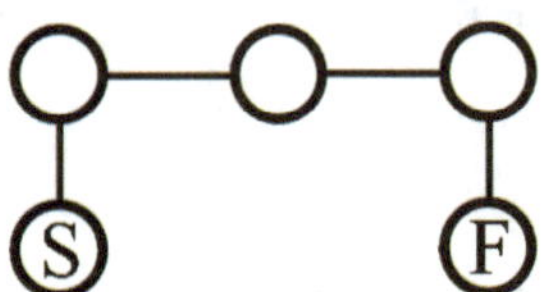

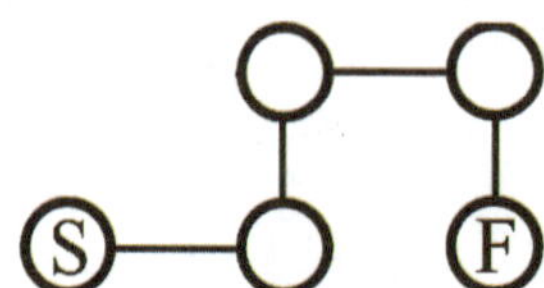

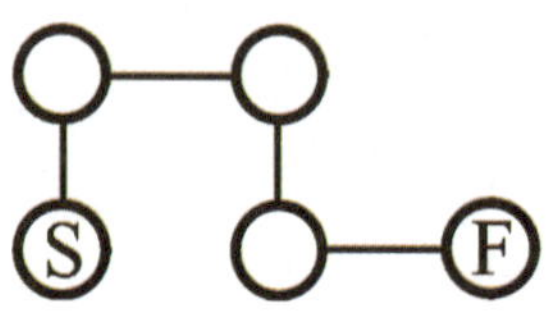

18. (D) 15

The pirate captain was number eight in line to climb the rope, so 7 pirates were ahead of him. Since there were exactly as many pirates behind the captain as ahead of the captain, there were also 7 pirates behind the captain. The total number of pirates to climb the rope was the 7 pirates ahead of the captain, the captain, and the 7 pirates behind the captain, or $7+1+7=15$ pirates.

19. (C) 18

We make a list of the possible numbers of mice caught each day. The numbers all have to be even, because the number of mice caught on day three is double the amount of mice caught on day one. Beginning with 0 mice caught, on each of the following days the number of mice caught is two more than on the previous day. We then take three consecutive numbers and check to see if the last number is twice the first number. The picture below illustrates such a sequence. The numbers in the box add up to $4+6+8=18$.

0 2 | 4 6 8 | 10

20. (D) 8

The sum in the row and the sum in the column are both made up of three numbers. There are four odd numbers and one even number. If 8, the only even number, is a number in only one of the sums, the sums cannot be equal because one will be odd and the other even (three odd numbers make an odd sum, but two odd numbers plus one even number make an even sum). So, 8 must be part of both sums, and so must be the number in the central square.

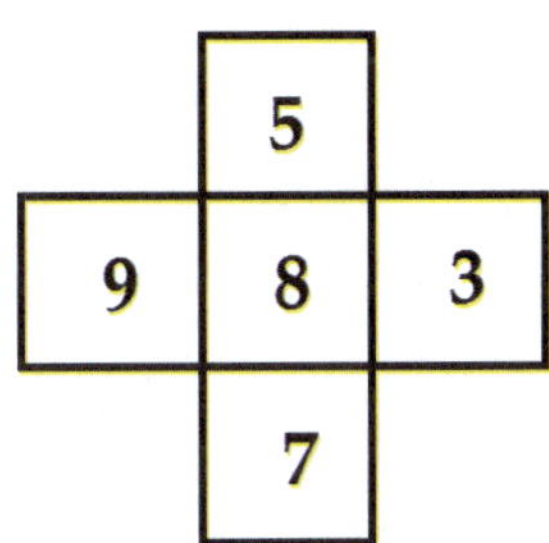

21. (D) 19 hens and 16 geese

Atos the dog and all the hens count for half of the 40 animals, which is 20, so there is 1 dog and 19 hens. The other 20 animals are all geese and ducks. For each duck there are 4 geese. Put them in a number of groups, each group with 1 duck and 4 geese. There are 4 such groups since $4\times 5=20$, there are 4 ducks and $4\times 4=16$ geese. So, grandmother has 1 dog, 19 hens, 4 ducks, and 16 geese.

22. (B) 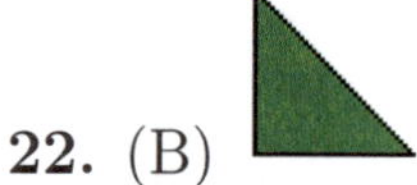

Note that the kangaroo sticker is the only sticker not visible on any of the faces, and all other stickers are visible. We rotate the die on the left as shown in the diagram below to obtain the die on the right. The blue circle is opposite of the red square and the yellow star is opposite the brown arrow. Therefore, the kangaroo is opposite the green triangle.

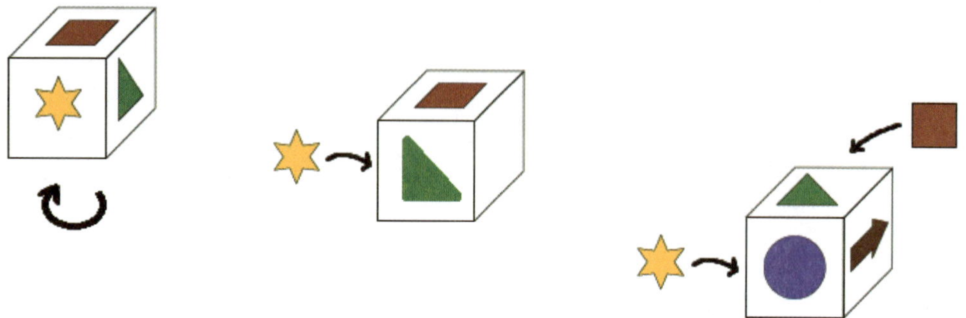

23. (C) Wanda bought the waffle

From the last statement we know that the order of the girls was Tara, then Sylvia, then Una. From the second statement we know Una was not last, so she must have been followed by Wanda, who was the last girl. Since Una did buy the cake, we have the following order:

- Tara
- Sylvia
- Una with the cake
- Wanda

The first girl did not buy the ice cream or waffle, so only the bun was left. Sylvia did not buy the waffle so she bought the ice cream. We have:

- Tara with the bun
- Sylvia with the ice cream
- Una with the cake
- Wanda with the waffle

24. (A) 13 hours 39 minutes

From 4:32 p.m. until 5:00 p.m. is 28 minutes. From 5:00 p.m. to midnight is 7 hours, and from midnight to 6:11 a.m. is 6 hours and 11 minutes. We add up the hours to obtain $6 + 7 = 13$ hours, and we add up the minutes to obtain $28 + 11 = 39$ minutes. The total travel time was 13 hours and 39 minutes.

Solutions for Year 2017

1. (D) David
Follow the fishing line from the fish to see that it was David who caught it.

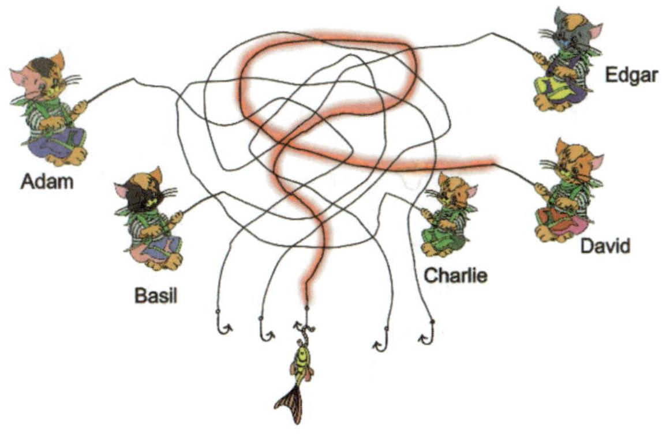

2. (C) 4
The 4 four stars that have only 5 points are circled in red
below.

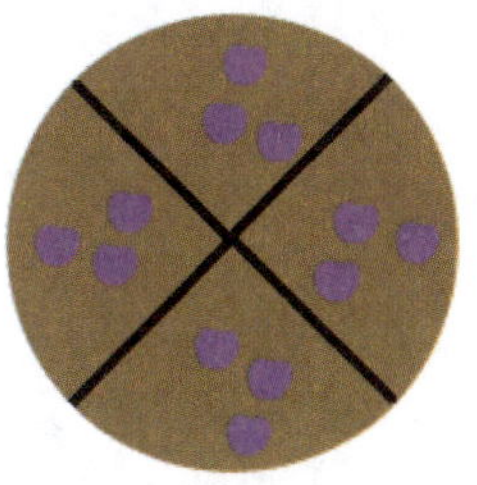

3. (B) 4
There are 4 children since the pie can be divided into 4 pieces with 3 cherries each.

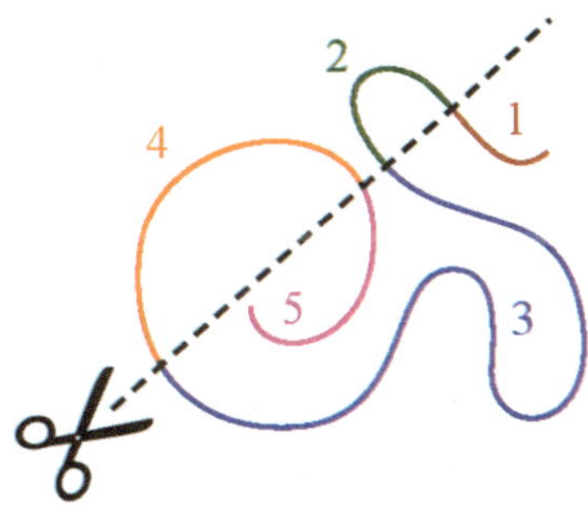

4. (A) 5
The scissors cut the rope into 5 parts as shown.

5. (A) 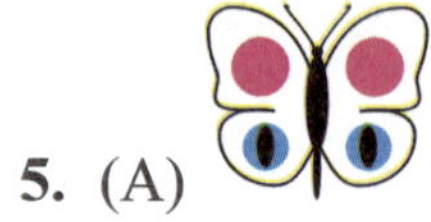

Ellen can only make butterfly (A) using all six stickers.

Here is why she cannot make the other butterflies:

(B)  Ellen does not have the small round black stickers.

(C) 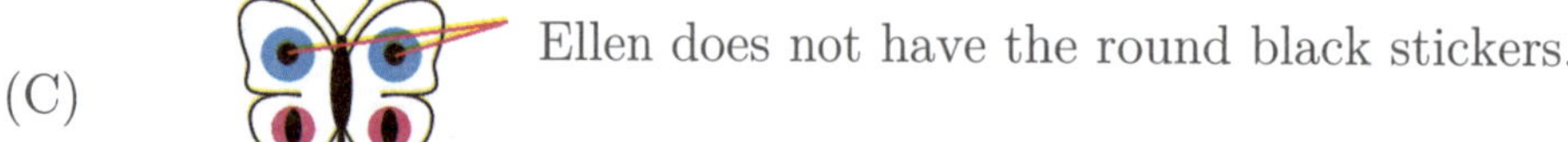 Ellen does not have the round black stickers.

(D) 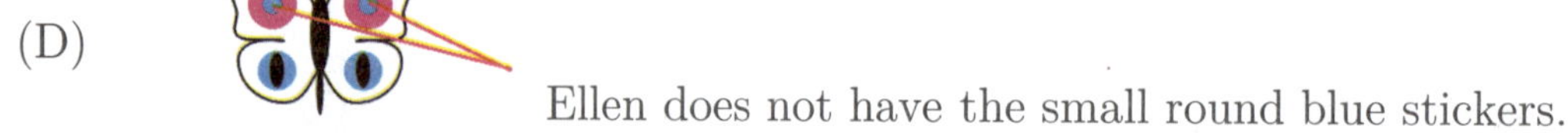 Ellen does not have the small round blue stickers.

(E)  Ellen does not have the round black stickers or the small blue stickers.

6. (A) 6

All the bricks are the same shape and size, so we can draw in the missing bricks and count them.

7. (E) 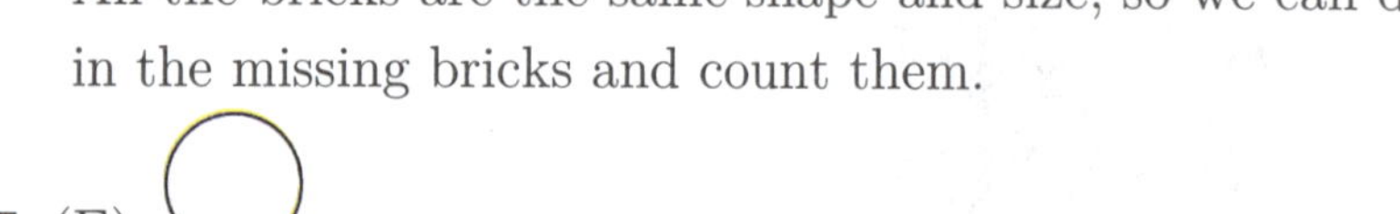

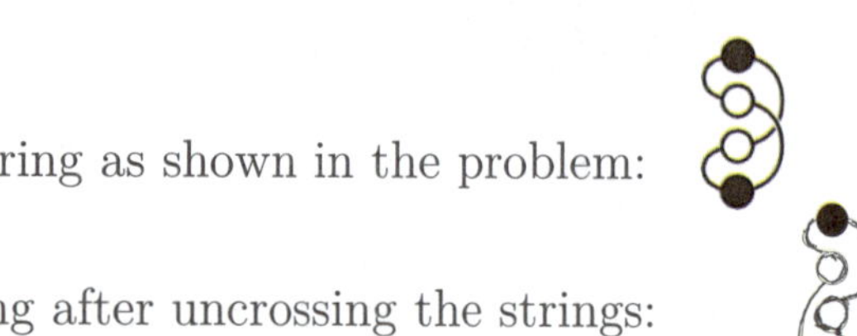

This was the string as shown in the problem:

This is the string after uncrossing the strings:

Slide the beads together to see that this is the order of the beads on the string: 

8. (C) 4

Notice that 4 is the only even number among the five numbers. If we use 4 on one side, then the sum with 4 is an odd number. On the other side of the equality we have to add two odd numbers, so their sum is an even number and the equality is incorrect. So, 4 cannot be used. When we use only the odd numbers, $1 + 7 = 5 + 3$ is an example of a correct equality. The numbers can be moved as long as each side adds up to 8.

$\boxed{1} + \boxed{7} = \boxed{5} + \boxed{3}$

9. (D) 12

There are three different objects for trading:

 sapphire  ruby  flower

The problem states that

 = 

and that

 =   

Knowing this, we can see that:

  = 

Since we can trade one ruby for three sapphires, and the three sapphires for six flowers, we can trade two rubies for twelve flowers. Notice that $12 = 2 \times 3 \times 2$, which represents 2 rubies, 3 sapphires for one ruby, and 2 flowers for one sapphire.

10. (A) 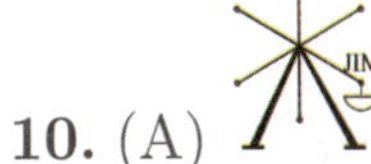

Ben moved a distance of 4 seats clockwise in order to be where Jim was. At the same time Jim also had to move the distance of 4 seats clockwise. Answer (A) shows the place where Jim ended up.

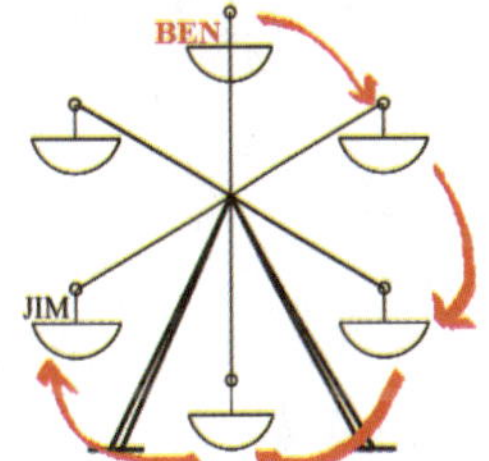

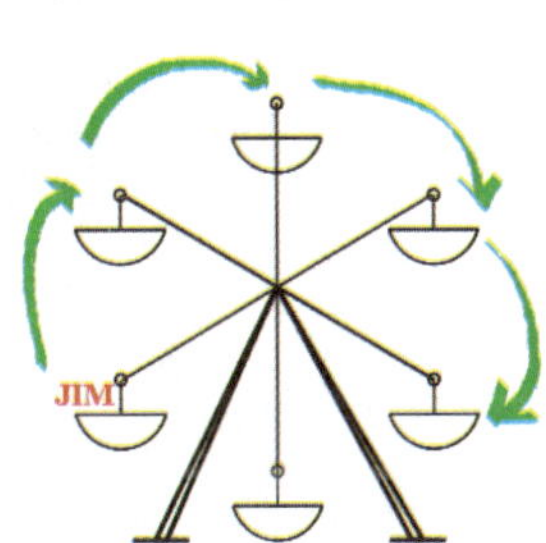

11. (D) 11

There are 5 triangles on the roof. In the picture here, four of them are the red, green, blue, and purple triangles, and the fifth is the entire triangular roof shape consisting of all four colored triangles. There are 6 triangles on the side of the house (below the roof). 3 are those colored brown, pink, and black, and the other 3 are made by putting together the pink triangle with the black triangle, the pink triangle with the brown triangle, and the gray quadrilateral with the brown triangle. Together there are 11 triangles in the picture.

12. (E) 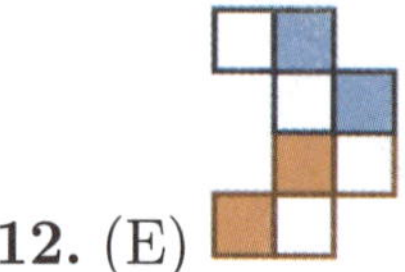

Notice that after four turns the shape will be in its original position. After two additional turns it will be in the same position as after two turns.

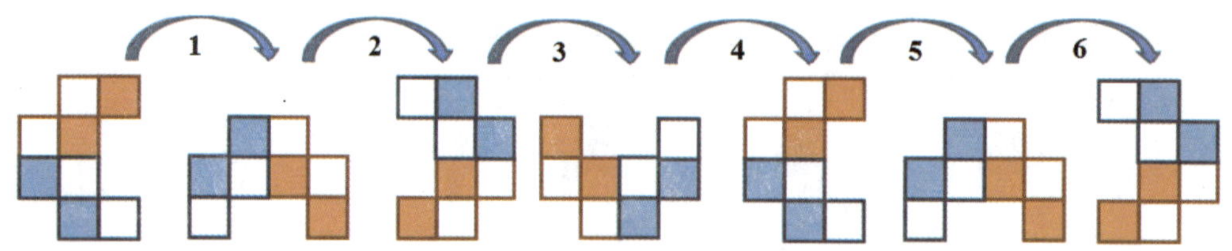

13. (D) 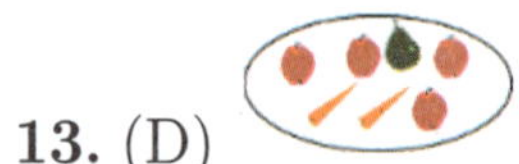

Look at each of the five pictures and write down the number of apples, pears, and carrots.

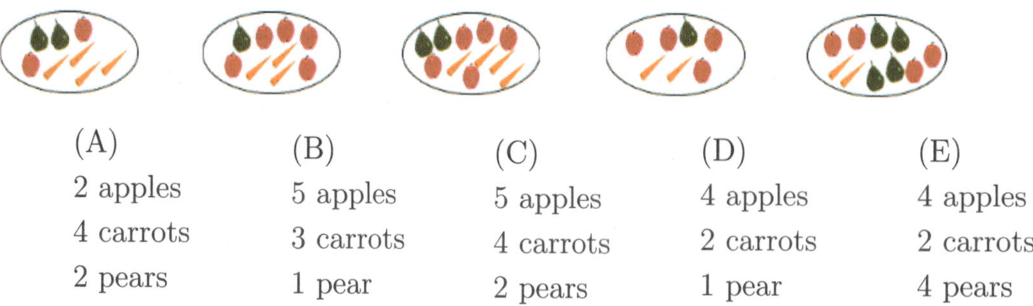

There are twice as many apples as carrots in pictures (D) and (E) since 4 is twice as much as 2. There are twice as many carrots as pears in pictures (A), (C), and (D) because 4 is twice as much as 2 and 2 is twice as much as 1. So, the only picture that has both twice as many apples as carrots and twice as many carrots as pears is (D).

14. (A) 2

Start with the fact there are 11 people in line. There are 7 people in front of Brian, so he is eighth in line. William is behind Brian, so William is ninth. Since there are 11 people total and 11 – 9 = 2, there are 2 more people behind William.

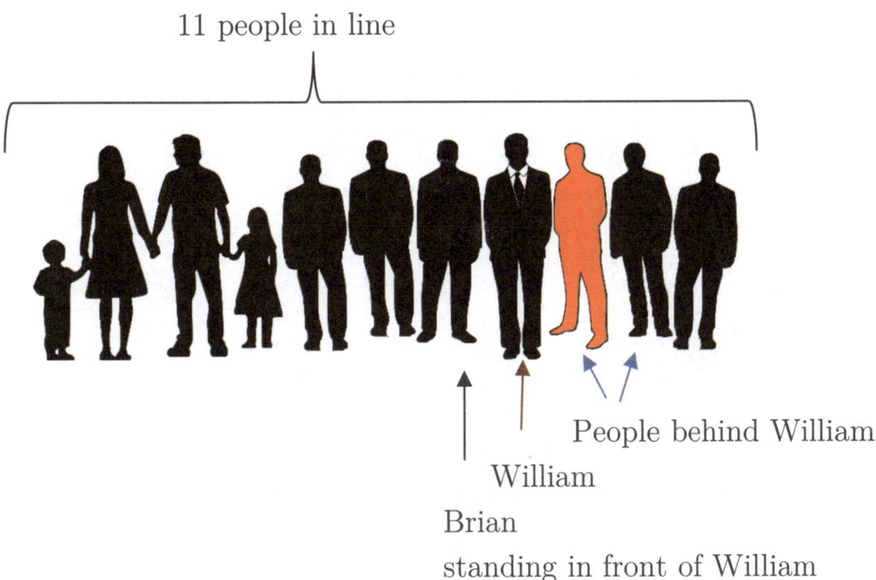

15. (C) 2012

The Roman numeral XX stands for 10 + 10 = 20. XV stands for 10 + 5 = 15. It is 2017 now and the 20th Math Kangaroo is taking place. The 15th Math Kangaroo was 5 years ago, which was in 2017 – 5 = 2012.

16. (B) 4

In order to make the three crowns Liz will need 3 crosses, 12 dots, and 3 lines.

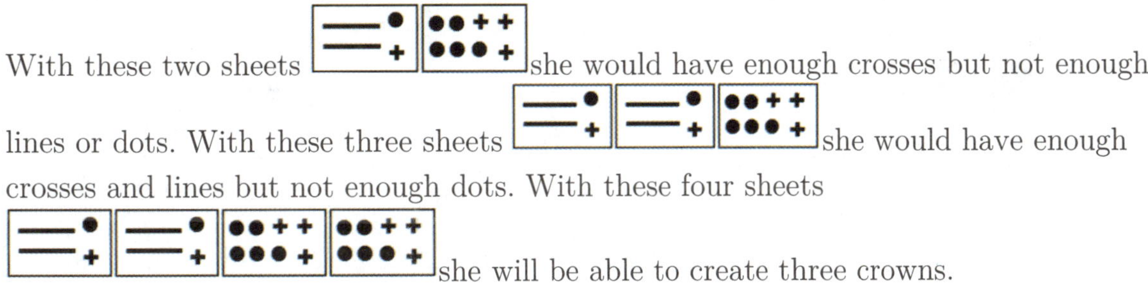

With these two sheets she would have enough crosses but not enough lines or dots. With these three sheets she would have enough crosses and lines but not enough dots. With these four sheets she will be able to create three crowns.

17. (B) 11

The square shows the sums of the number above the column and the number in front of the row. In the second row, we can figure out that the number in front of the row is 4, because we need to add 4 to 10 to get 14. The question mark should be replaced with number 11 as it is the sum of 7 and 4.

+	10	7
5	15	12
4	14	11

18. (C) 2

At Old McDonald's Barn there is 1 horse, 2 cows, and 3 pigs as shown in the picture below. There are 6 animals total, and 6 is not twice as much as 2, so we need to add more cows.

If one cow is added to the barn, the barn would have 1 horse, 3 cows, and 3 pigs. The number of all the animals would be 7. The number of cows would be 3. 7 is not twice as much as 3.

If 2 cows are added to the barn, the barn would have 1 horse, 4 cows, and 3 pigs. The number of all the animals would be 8. The number of cows would be 4. 8 is twice as much as 4.

So, Old McDonald's Barn needs 2 more cows so that the number of all the animals is twice the number of cows.

19. (C) 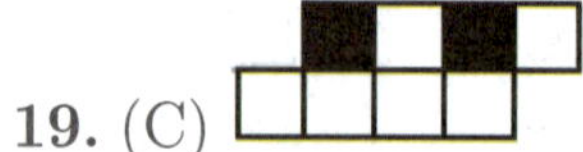

Only shape (C) can be made using both pieces, as shown in the picture on the right. All the remaining shapes can contain at most one such piece. Notice that each black cell must have at least 2 white neighbors with a common edge. This eliminates options (A) (it also has 3 black cells), (B), and (E). (D) can also be eliminated. The black cell in the upper row is adjacent only to 2 white cells, so these cells came from one piece. The second black cell (in the lower row) is from the second piece and it is adjacent to 3 white cells. Two of these 3 white cells are from the first piece, so only one white cell from the second piece is adjacent to the second black cell but two such white cells are needed, so (D) can't be made.

20. (E) 9

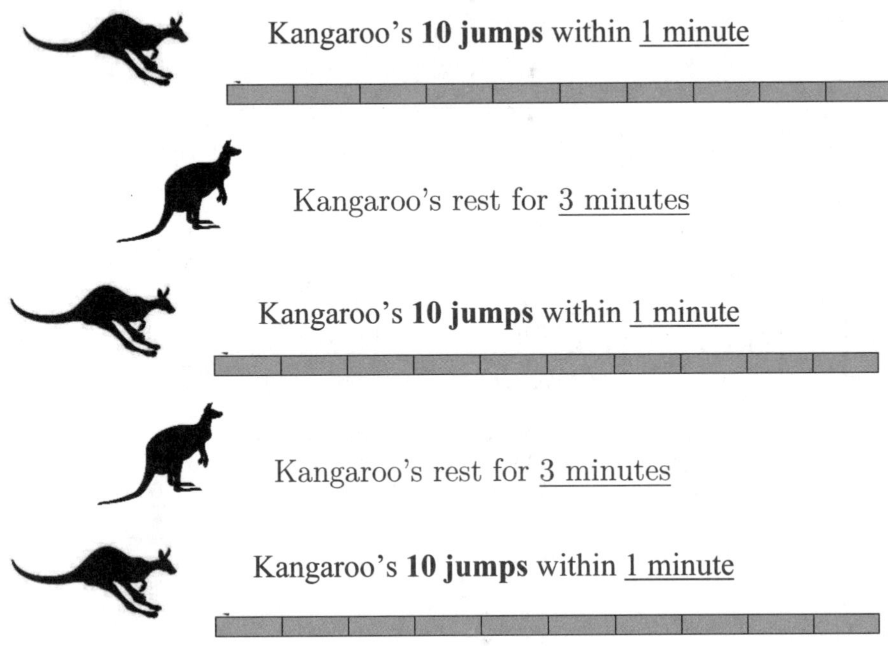

In order for the kangaroo to make 30 jumps he must first make 10 jumps, then 10 jumps and again 10 jumps. Between the three sets of 10 jumps (1 minute each) there are 2 rests (3 minutes each). Adding the minutes together, the kangaroo would complete 30 jumps in $1 + 3 + 1 + 3 + 1 = 9$ minutes.

21. (E)

The picture on the stamp must be the reverse or mirror image of the picture given. Notice that the picture of the castle has the tower on the left side with no windows to the left of it. The stamp would have to have the tower on the right side with no windows to the right of it. Only stamp (B) and stamp (E) are like this. Also, the picture shows one side of the castle with a door on the left and two windows on the right. The stamp would have to have a door on the right and two windows on the left. Only stamp (E) has this.

22. (D) GAG

There are exactly two keys which start with the same digit (4) and two locks which start with the same letter (D), so digit 4 corresponds to the letter D. The first lock (ADA) has the letter D in the middle and the only key with the digit 4 in the middle is the first key (141), so the first key (141) fits the first lock (ADA) and the digit 1 corresponds to the letter A. The two middle keys fit the two middle locks (471 with DGA, and 417 with DAG), and the digit 7 corresponds to the letter G. Finally, the last key (717) fits the last lock which has GAG written on it.

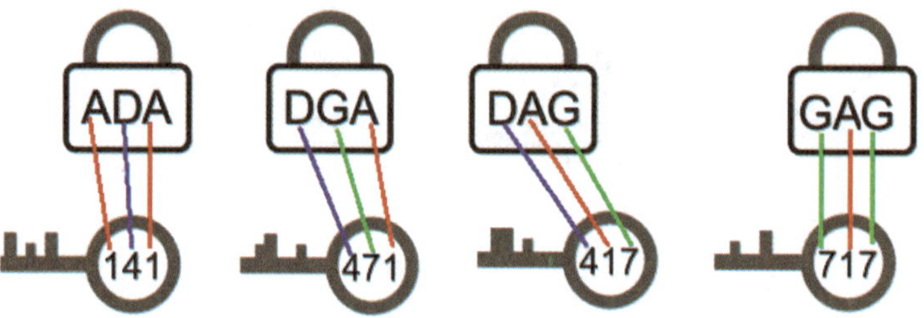

23. (B)

must be placed in the top row since it is above, and is in the bottom row. Given the third statement, we know that the bottom row will look like this:

.

Now we know that is on the far left of the top row, and from the first statement given we know that the top row will look like this: . This means that

is in the gray cubby.

24. (E) 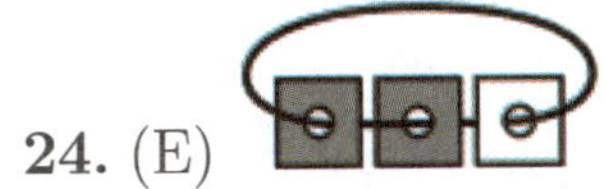

We have to focus on the way the rope shows on either the right or the left side of each card. In the original rope shown in the picture, the cards will all show the same color if they are flipped to the same side.

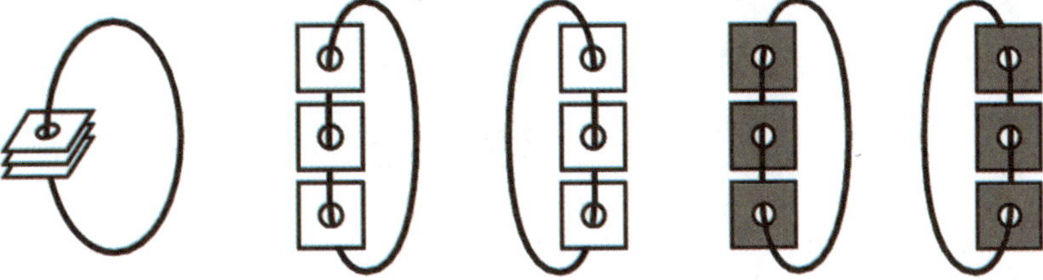

Because each card is white on one side and gray on the other, flipping a card will change its color. Use your imagination or sketch the rope with the cards to see how the cards' color would change if the cards are flipped. By flipping all the cards to one side we can see which of the choices (A) through (E) can be obtained without untying the rope. Notice that by flipping the third card in answer (E) the rope would be showing on the left side of each card and all the cards would be gray.

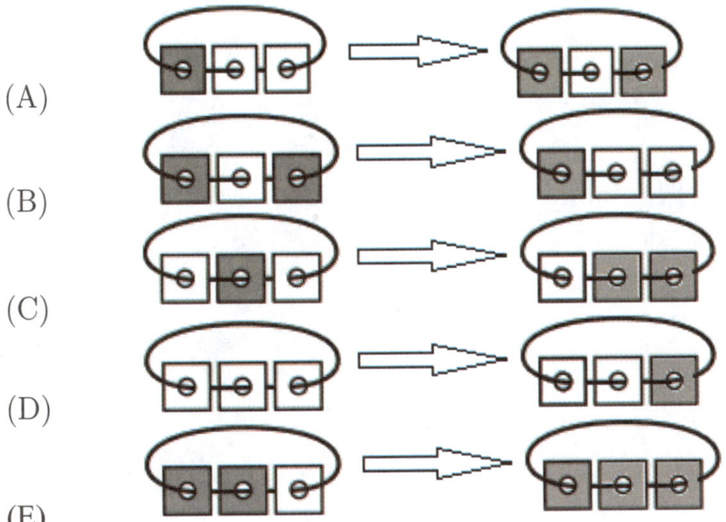

Solutions for Year 2019

1. (D) 

Remember that the number 7 does not count as less than 7.

In (A) 

the number 8 is more than 7. The number 7 is not less than 7.

In (B) 

the number 9 is more than 7.

In (C) 

the number 7 is not less than 7.

In (D) none of the numbers is less than 7.

2. (C) 

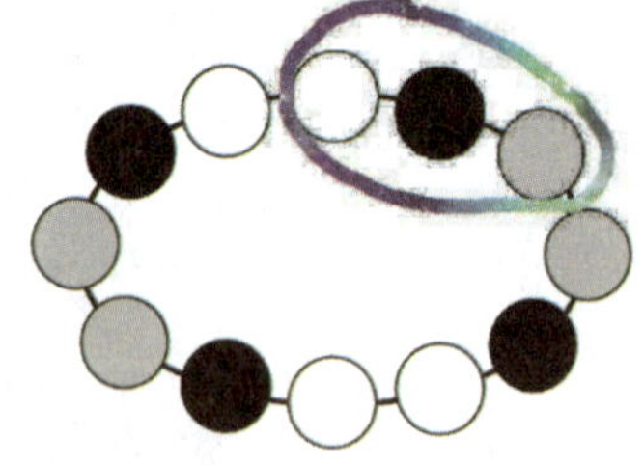

We do not have a place in this necklace where a white bead is between two shaded beads, and we do not have a place where any bead is between two beads of the same color.

3. (C) 8 kilograms

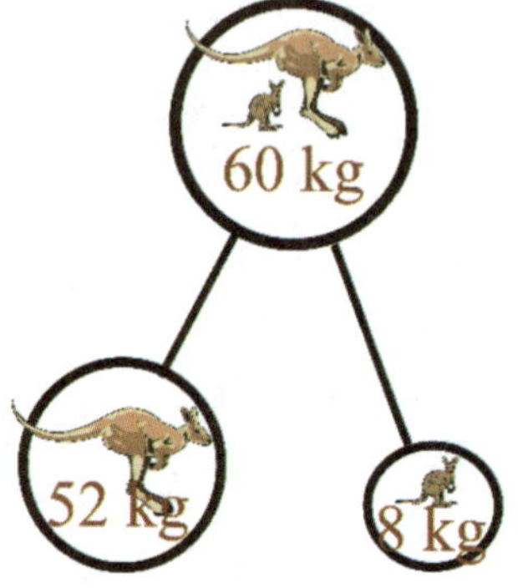

60 kilograms is the weight of Mom and Jumper. Subtract 52 kilograms from 60 kilograms to get the weight of Jumper.
$60 - 52 = 8$

4. (E)

The question does not say that the piece was not rotated after cutting, but we need to pay attention to which way the triangles face. There is no place where a triangle faces the way shown and is next to a circle or a star.

Piece (E) could have been cut out from any of the places outlined in color the picture.

5. (B) 3

Ignore the dog. The picture shows 12 people in line and helps us solve the problem.

6. (C) 5

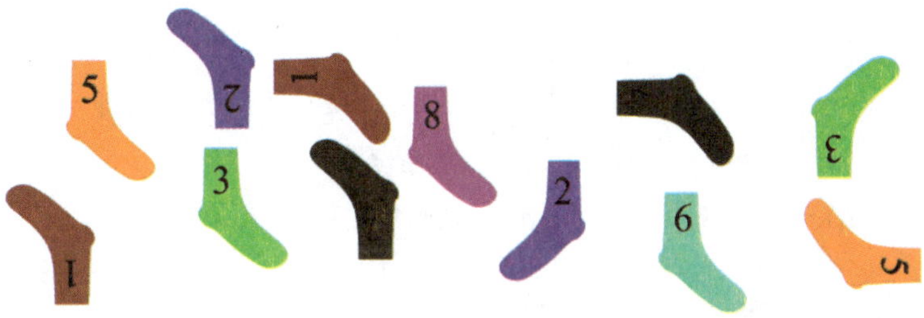

The socks with numbers 1, 2, 3, 5, and 7 can be paired up. This makes 5 pairs. The number 4 is missing and 6 and 8 don't have matches.

7. (A) 9

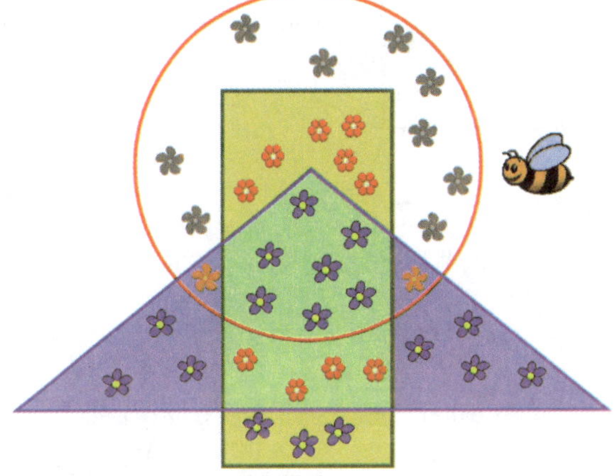

Only count the 6 flowers in the top part of the rectangle and the 3 flowers in the bottom part. $6 + 3 = 9$. Do not count the flowers in the part where the big blue triangle and the rectangle overlap.

8. (C) 7 cents

If two apples cost 6 cents, one apple costs 3 cents, because $3 + 3 = 6$.

If two pears cost 8 cents, one pear costs 4 cents, because $4 + 4 = 8$.

So, one apple and one pear cost 3 cents + 4 cents = 7 cents.

9. (E) 4 and 5

The only way the mouse can get to the cheese is through gate 4 or gate 5, so these need to be closed.

10. (B) 3

The picture below shows the edges of the paper in red so we can better see how it is folded. If we unfold the paper after cutting, we see that there were two lines where the paper was cut, dividing it into three pieces.

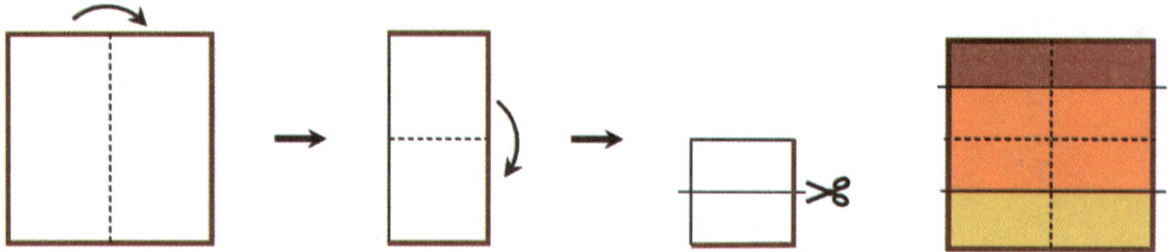

11. (A) 5-2-3-1-4

Each piece is square, and there is only one piece where you can see the whole square, so you know this one is on top and needs to be taken off first. This should allow you to see all of the next piece, and so on.

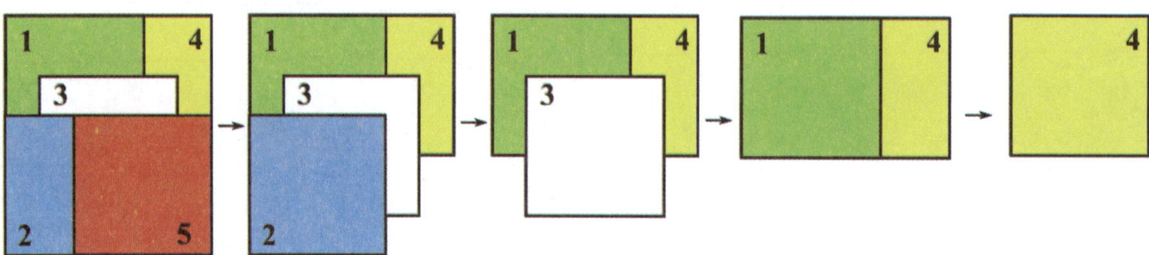

12. (E) 6

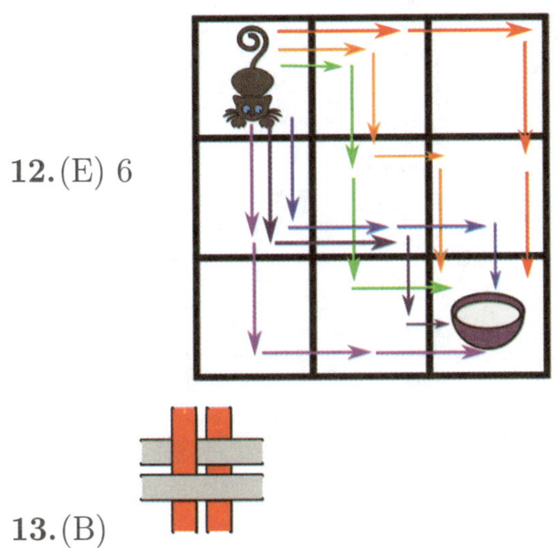

13. (B)

When looking from the front, you can see both red strips over the bottom gray strip, so looking from the back both must be covered. On the top part, the strip that is showing must be partly hidden and the part that is hidden must show. However, you are looking from the other side, so the strip that is showing will still be on the left.

14. (B) 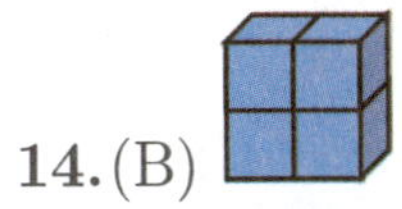

Each cube has 6 faces, but the way they are glued together covers up some of the faces. Count the number of faces that are not covered and find the figure with the fewest.

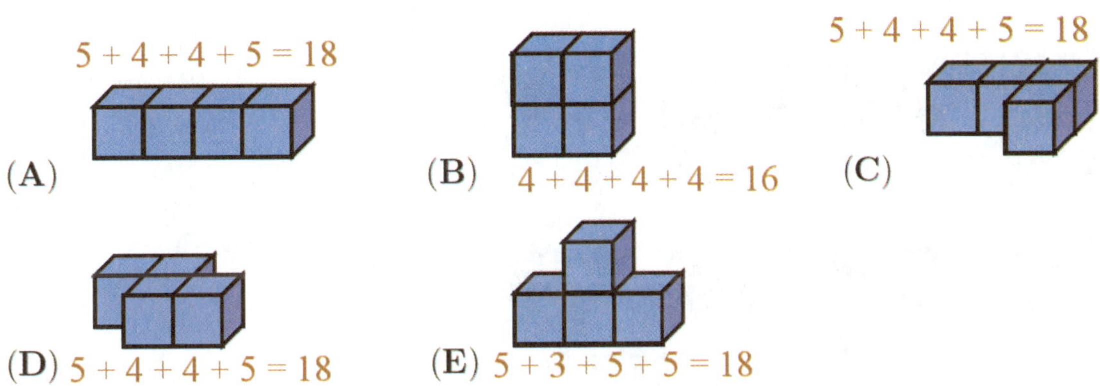

15. (E) 12 m

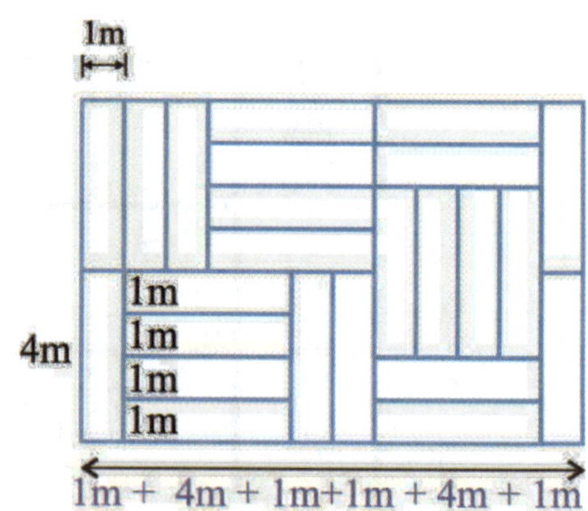

We can tell from the way the pattern is laid out that the long side of each rectangle measures as much as four short sides. 4×1 m $= 4$ m.
Add the lengths together along the bottom edge to find the length.

16. (D) 5 hours

First, find the time the whole trip takes. From 6:00 in the morning until noon is 6 hours, and then from noon to 11:00 at night it's 11 hours. 6 hours + 11 hours = 17 hours.
Next, find the total travel times marked on the station.
$2 + 3 + 7 = 12$.
Subtract to find the missing number. $17 - 12 = 5$.

17. (D) 24

$8 + 8 = 16$ sheep

8 cows

There are 8 more sheep than cows. In order for the number of cows to be equal to half the number of sheep, the 8 sheep need to be half of the sheep; the other 8 sheep have 8 cows to match the number. So, there are $8 + 8 = 16$ sheep and 8 cows. $16 + 8 = 24$

18. (A) 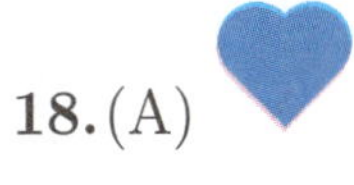

19. (D) 4

If all the camels were dromedary, there would be 10 humps. For each Bactrian camel, there is one hump more than 10. 4 – 10 = 4, so there are 4 Bactrian camels.

20. (B) 2

We need to find three different whole numbers that add up to 7. The middle number will be how many nuts Elli collected. Trying with the smallest numbers possible, we quickly see that the only numbers that work are 1, 2, and 4. The middle number is 2.

Anni	Elli	Asia	
1	2	3	= 6
1	**2**	**4**	**= 7**
1	3	4	= 8
2	3	4	= 9

21. (C) 50 cm

Since the upper tip is 80 cm from the ground and the lower tip 20 cm from the ground, the flagpole is 80 cm – 20 cm = 60 cm tall. Half that height is 30 cm. Add the distance from the ground to the bottom of the flagpole and half the height of the flagpole to get the height of the sandcastle.
20 cm + 30 cm = 50 cm.

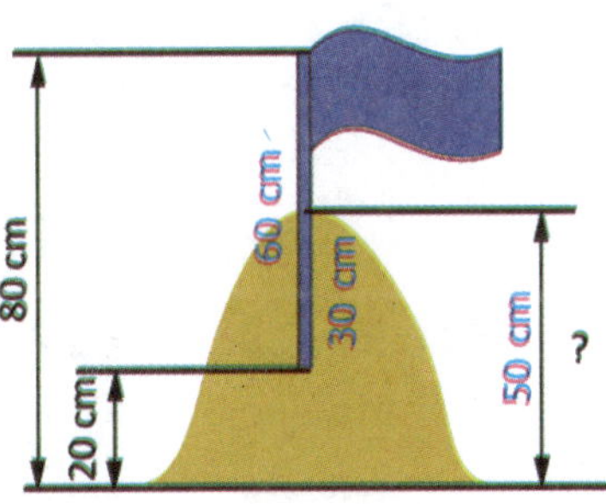

22. (D) 

We need to make the changes in order. At first, the strip looks like this:

After Ani makes all the black squares into white squares, it looks like this:

After Bob makes all the gray squares into black squares, it looks like this:

After Chris makes all the white squares into gray squares, it looks like this:

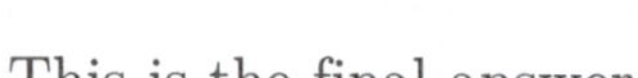

This is the final answer.

23. (A) 14

The only two 4-cell squares that give a sum greater than 63 are 13, 14, 18, and 19 (the sum is 64) and 14, 15, 19, and 20 (the sum is 68). The 4-cell square with the numbers 12, 13, 17, and 19 has a sum of only 60, and any other squares will have sums smaller than that.

1	2	3	4	5
6	7	8	9	10
11	12	13	14	15
16	17	18	19	20

Both 14 and 19 appear in both squares that work, but only 14 is listed as one of the answers. (Since the questions asks which of the numbers listed in the answers must be in the square, 19 would not be a correct answer.)

24. (C) 8

Putting a white token in the machine gives one more token than before. Putting a red token in the machine gives two more tokens than before. To get the smallest answer, we need to think about Amalia avoiding putting red tokens into the machine.
There is no way to avoid changing at least one red token into three white ones. However, Amalia has one white token to start with, and then changing one red token gives her more white tokens to change.
So, Amalia needs to make one change that will give her two extra tokens, but the other two changes can each give only one extra token. That means she ends up with $2 + 1 + 1 = 4$ extra tokens. This added to the original 4 makes 8.

Solutions for Year 2021

1. (E)

Count the number of straight sticks in each figure. Only figure (E) uses 3 sticks. The rest use 4 sticks.

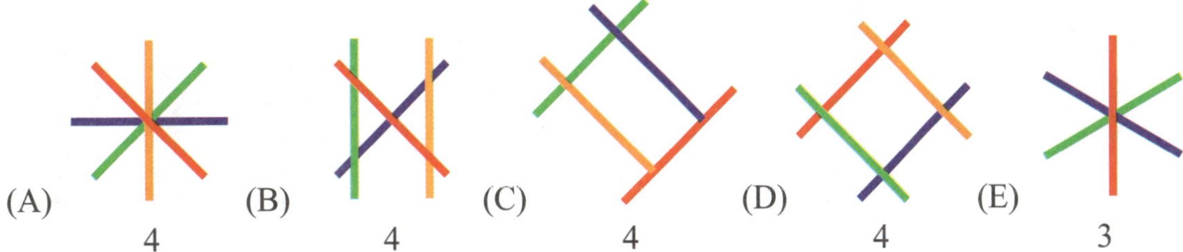

2. (B) 5

The scale on the right side shows 12 intervals from 0 to 12, so each mark stands for 1 unit. The taller mushroom measures 11. The shorter mushroom measures 6. The difference is 11 – 6 = 5.

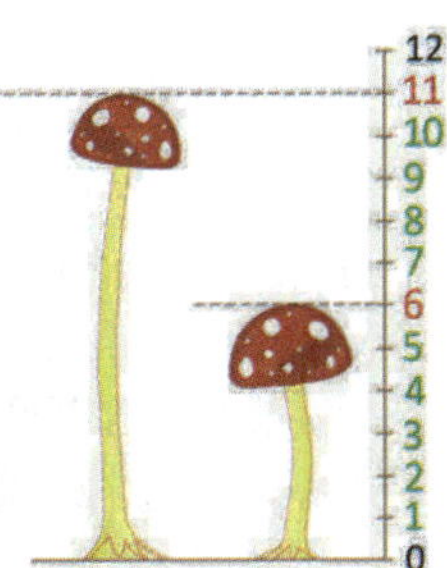

3. (A)

Each path is made up of bold segments of the same length. Count the number of segments in each path to find the longest one.

(A) 15 (B) 13 (C) 12 (D) 12 (E) 11

4. (D) D

Each piece of paper is a square identical to the one on top of the pile. The diagram on the right shows the outlines of the squares. Only a hole at D is within the outlines of all four squares.

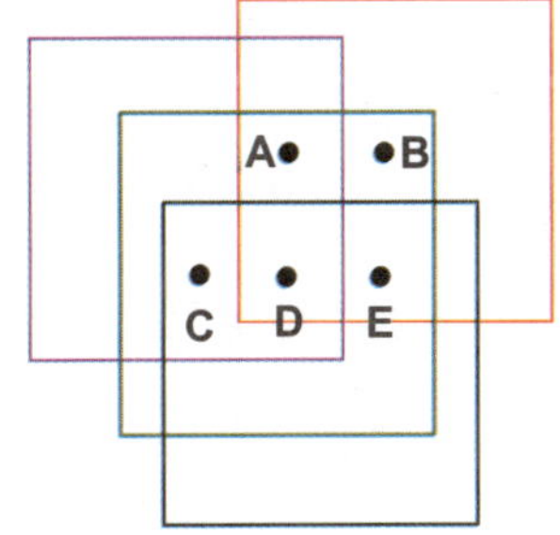

5. (A)

In a mirror, all the digits are reversed, so (D) and (E) cannot be the right answer. Also, the number is in reverse order, so the order of the digits from left to right is now 1-2-0-2.

6. (D) silver tower

The first statement tells us that the green tower is taller than the pink and red towers. The second statement tells us that the silver tower is even taller. So, the silver tower is the tallest.

7. (E) 6

If a child is facing backward in the picture, his or her right hand is on the right. If the child is facing forward, his or her right hand is on the left. Circle all the right hands and see how many are holding another child's hand.

Another way to solve: There are seven children, so there are seven right hands. Of the seven children, only the girl on the far left is not holding another child's hand with her right hand. So, the remaining six children are holding another child's hand with their right hand.

8. (B) 

Only the stars in picture B don't include any numbers 3 or less. Check that $8 + 7 + 5 = 15 + 5 = 20$.

9. (D) 12

It's easier to count the number of cuts than to follow the winding string. There is always one more piece than the number of cuts. For example, one cut divides a string into two pieces and two cuts divide it into three pieces. There are 11 cuts, marked with dots on this picture, so there are 12 pieces of ribbon.

10. (D) D

The distance around the wall is 4m+1m+5m+2m+3m=5m+5m+5m=15m. The cat needs to walk another 20 m – 15 m = 5m after she makes it all the way around and gets back to point B. Walking to point C adds another 4 meters, for a total of 19 meters, and another 1 meter will bring her to point D.

11. (C) 6

There are three types of flowers: white, gray, and black. We need to figure out how many of each flower needs to be added to either pot and then add those numbers together.

There are 2 white flowers in the first pot and 4 in the second pot. Julia will have the same number of white flowers in each pot when she adds 4 – 2 = **2** white flowers to the first pot.

There are four gray flowers in the first pot and one in the second pot. She will have the same number of gray flowers in each pot when she adds 4 – 1 = **3** gray flowers to the second pot.

Finally, there are two black flowers in the first pot and three in the second pot. Julia will have the same number of black flowers in each pot when she adds 3 – 2 = 1 black flower to the first pot.

Altogether, Julia needs to add 2 + 3 + 1 = 6 flowers to the pots. This is the smallest number of flowers she needs to buy.

12. (E) MATH

The letter tells us the column and the number tells us the row. B3 stands for M, B2 for A, C4 for T, and D2 for H. This spells out "MATH."

13. (A) 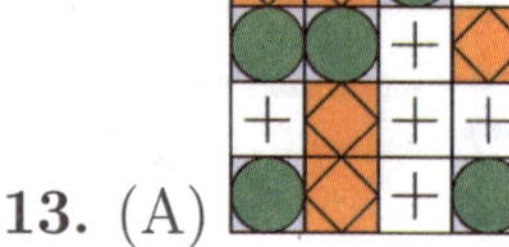

Both pieces will need to be rotated to fit together and to match one of pictures shown.

14. (C) 8

Angela earned $2 \times 9 = 18$ points by scoring 9 goals. Julie earned $2 \times 5 = 10$ points by scoring 5 goals. $18 - 10 = 8$, so Angela earned 8 points more than Julie.

15. (B) 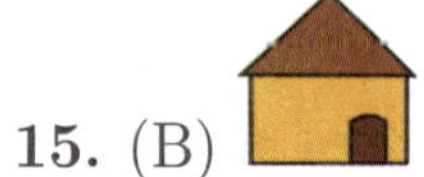

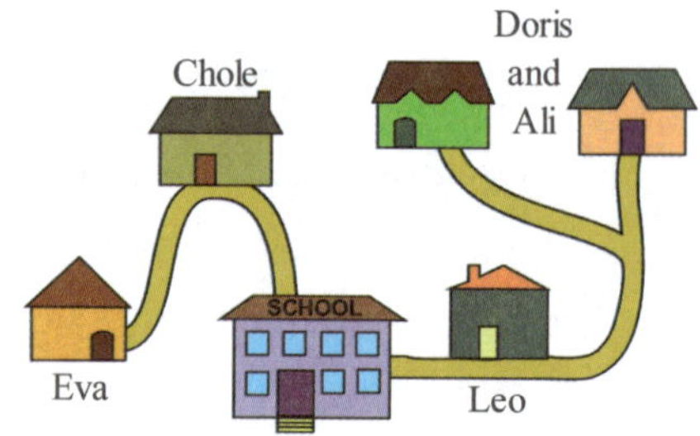

Doris and Ali both walk past Leo's house, so Leo's house must be the gray house with the pink roof on the right with the path splitting in two behind it. So, Eva's and Chole's houses are on the left side of the school. Eva walks past Chole's house, so her house is the one further away.

16. (D) 10

There is a total of $10 \times 2 = 20$ leaves on the two branches. The leaves the kangaroo ate from the second branch matches exactly the leaves that were left on the first branch. For example, if it ate 2 leaves from the first branch, it ate $10 - 2 = 8$ leaves from the second branch. If it ate 5 leaves from the first branch, it ate $10 - 5 = 5$ leaves from the second branch. So, no matter how many leaves it ate from the first branch, the total number of the leaves it ate is 10. This means there are $20 - 10 = 10$ leaves left.

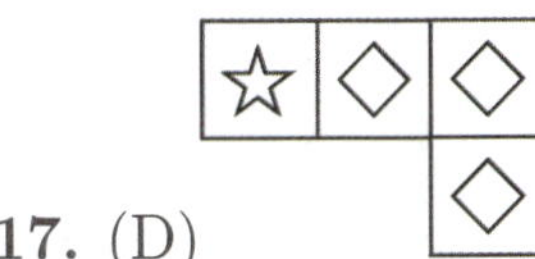

17. (D)

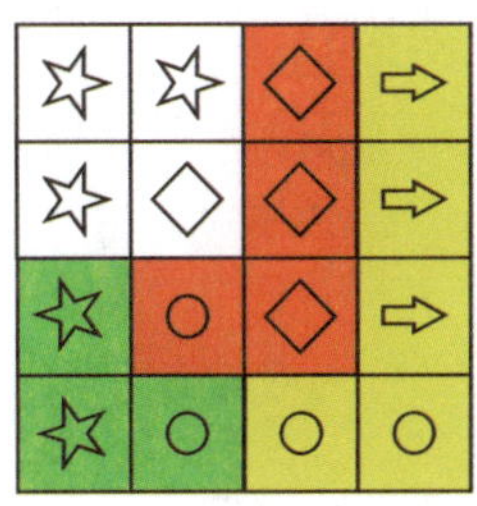

All five answers listed can be found in the square, so we need to determine which four were actually used to construct it. Some of the pieces need to be rotated. Piece (E) was used as it's the only one that has the arrows. Then, we need two more circles, so both pieces (B) and (C) were used. The remaining piece needed to make the square is (A). This leaves piece (D) as the one that was not used.

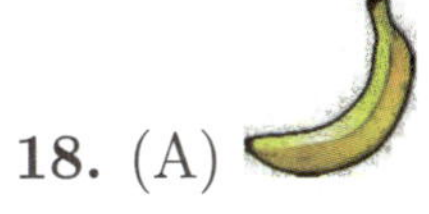

18. (A)

From the 4 apples the witch starts with, she makes one banana and has one apple left. From the 5 bananas she has, she makes one apple and has two bananas left.

At this point, she has two apples and three bananas. She turns the three bananas into one apple and has three apples.

Finally, she turns the three apples into a banana, which is what she ends up with.

19. (C) only 6

The sum of all 5 numbers is $2 + 3 + 4 + 5 + 6 = 20$. If the cards are placed in two boxes with the same sum, each box needs to have a sum of $20 \div 2 = 10$. The only way to make a sum of 10 using 4 and any of the remaining numbers is to add the number 6. Check that the other three numbers also add up to 10: $2 + 3 + 5 = 10$.

20. (C) 

Each of the gears will move the same number of teeth, but in opposite directions. The smaller gear has 8 teeth and the larger gear has 16, so the larger gear will move half a turn while the small gear move a full turn. The black tooth on the larger gear will be halfway across the gear from where it started.

21. (B) 6

Each of the three girls danced with each of the two boys, so $2 \times 3 = 6$ different pairs danced. Since each dance took one minute and none of them took place during the same time, the dancing lasted 6 minutes.

22. (C) 3

The plate has three large star cookies, four small star cookies, three moon-shaped cookies, two triangle cookies, and two hearts. There is only one moon-shaped cookie on the tray, so three trays are needed to provide enough moon-shaped cookies. This also gives more than enough cookies of the other shapes.

23. (E) 24

Kangie eats 2 apples a day on three of the days of the week, so he eats $2 \times 3 = 6$ apples a week. He eats 3 mangoes on two days of the week, so he eats $3 \times 2 = 6$ mangoes a week. So, he eats $6 + 6 = 12$ pieces of fruit in one week and $12 \times 2 = 24$ pieces of fruit in two weeks.

24. (C) 3

The four toys other than the puzzle are placed in the following order from bottom to top: blocks, ball, game, car. The ball and the game need to be next to each other, but the puzzle could be between the blocks and ball, between the game and the car, or at the very bottom or the very top shelf. All five shelves are filled, and the puzzle is either one of two toys below the ball and game, or one of two toys above them. This means that it could be on shelf 1, 2, 4, or 5, but it is not on shelf 3.

Solutions for Year 2023

1. (D) 8
Be sure to count the outer circles as well as the inner circles. The head is 1 circle, the mouth is 1 more, and the eyes are 2 more. The ears are each made up of 2 circles, so there are 4 circles there. Altogether, there are 8 circles.

2. (B)
The cube on the top left is red and the one on the top right is blue. The top row does not have a cube in the middle, so the color we see is yellow. Therefore, from left to right, we see red, yellow, and blue. Even if we rotate the view, we need to have a yellow cube in the middle with red and blue on the sides.

3. (A)
Notice that in each bowl there is a ball with numbers 9, 7, and 4. So, the sum depends only on the fourth number and is the largest in (A).

We can also find the totals for each bowl. To make the adding easier, notice that $7+4+9 = 20$.

(A) $8 + 7 + 4 + 9 = 28$
(B) $4 + 6 + 7 + 9 = 26$
(C) $7 + 9 + 4 + 7 = 27$
(D) $9 + 4 + 4 + 7 = 24$
(E) $7 + 4 + 5 + 9 = 25$

The numbers in bowl (A) make the largest sum.

4. (A)

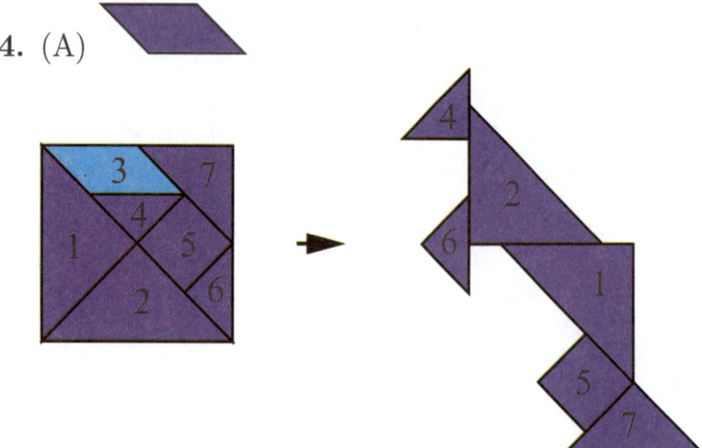

The kangaroo is made using five triangles and a square. So, Mr. Beaver did not use the piece we marked with the number 3 (the parallelogram).

5. (E)
My boat has more than 1 circle, so it's not (C).
Now we count how many triangles and squares are on the sails. In (A) there are 2 triangles and 1 square, which is 1 more triangle than the number of squares. (A) is not my boat.
In (B) there are 3 triangles and 2 squares, which is 1 more triangle than the number of squares. So, (B) is not my boat.
In (D) there is 1 triangle and 3 squares, so there are fewer triangles than squares. (D) is not my boat.
In (E) there are 4 triangles and 2 squares, which is 2 more triangles than squares, so this is my boat.

6. (C) 76
There are 7 big candles and 6 small candles on the cake. The 7 big candles stand for $7 \times 10 = 70$ years and the 6 small candles stand for 6 years. So, the grandfather is $70 + 6 = 76$ years old.

7. (B) 6
In total, there are 10 cars on the racetrack. We can see 4 cars that are not in the tunnel, so $10 - 4 = 6$ cars are in the tunnel.

8. (D) 14
Steven will stop each time it reaches a red dot, and then continue straight ahead. He will pass each of the crossings exactly twice. In total, he will stop 14 times.

9. (D) at D

From any point except D, the beaver can see more than 2 trees. From point D, the two big trees block all the rest from view.

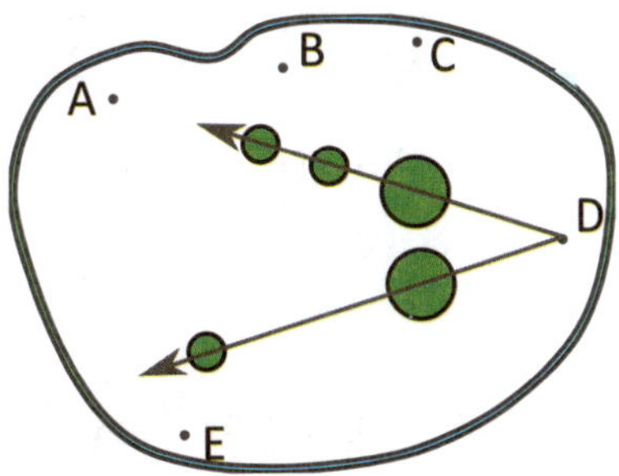

10. (C) 3

Half the number of squares is $24 \div 2 = 12$, and 9 squares are already colored. Therefore, $12 - 9 = 3$ more need to be colored.

Or:

Imagine bringing the colored parts as far down as they go, like in a Tetris game. Then, it's easy to see that coloring 3 more squares will make half of the figure colored.

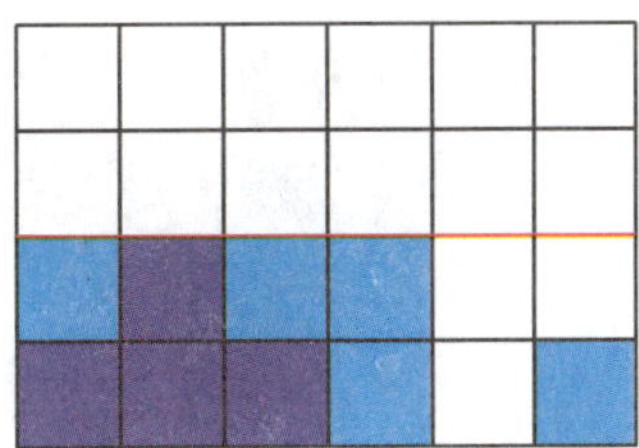

11. (C) 3

The two tokens with the numbers we know add up to $10 + 2 = 12$. The sum is 18, so the tokens with the question marks need to make $18 - 12 = 6$. They are both the same, so each one has the number 3 on it, because $3 + 3 = 6$.

12. (E) 13

The missing parts of the bee are 1 eye, 1 antenna, 1 wing, and 1 mouth. Altogether these use $1 + 3 + 4 + 5 = 13$ points.

13. (C) 12
We can color the squares and count how many are unpainted.

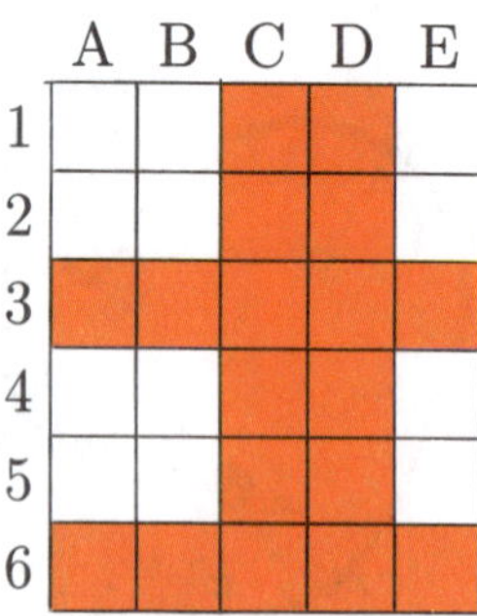

14. (B)
After unfolding, the two square holes will be closer together than the two round holes.

15. (D) 11

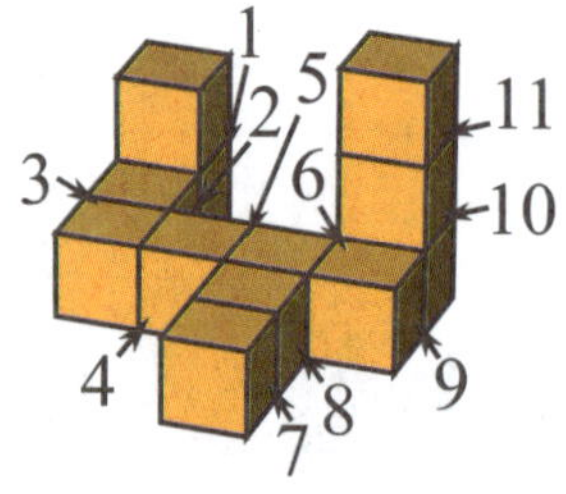

16. (A)

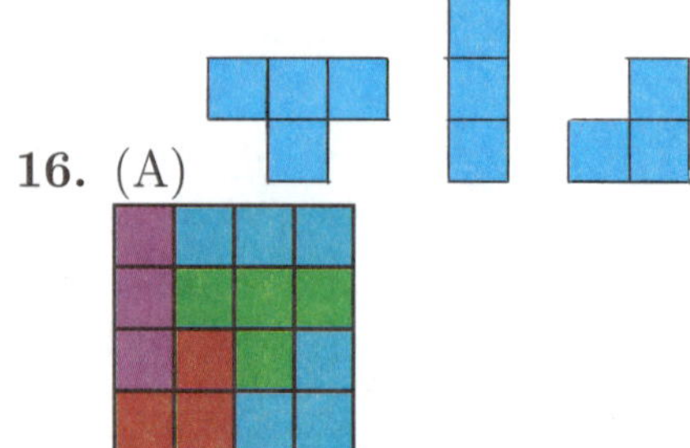

We can count the number of little white squares that need to be covered up to complete the puzzle. There are 10 white squares. Choice (A) is the only option where the pieces will cover a total of 10 small squares. Double-check that this can actually be done.

17. (A)
We only need to compare one piece of each option and notice that only (A) is correct.

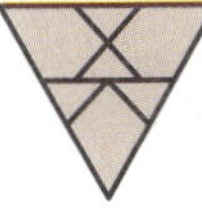

18. (C)
The five children are aged 8, 7, 6, 5, and 4. Vittorio is the youngest, so he is 4. Of the remaining children, if Lea is two years older than Jose and one year younger than Ali, then Jose and Ali are three years apart. Jose is 5 and Ali is 8. Lea is 7. This leaves 6 as Sarah's age.

19. (A) B and E
The whole circuit is $7+2+6+4+5 = 24$ km long. If the two distances are equal, each must be 12 km. So, let's look for a village that has another village 12 km away from it.
Let's start with village A and check going clockwise. Village B is 7 km away (too little), C is $7+2 = 9$ km away (also too little) and D is $7+2+6 = 15$ km away (too far). So A is not one of the villages.
Let's try B. C is 2 km away, D is $2+6 = 8$ km away, and E is $2+6+4 = 12$ km away. 12 km is the distance we are looking for, so the two villages are B and E. We can check that the other part for the road from E to B is also $5+7 = 12$ km long.
We can also check that none of the other cases, such as from C or D, do not give us a distance of 12 km to any village.

20. (B) 

To get from the entrance to the exit, Sam needs to at least visit the rooms numbered below. She will have to visit them in that order. She will see the shark in room 8, the rhinoceros in room 10, and the from in room 14.

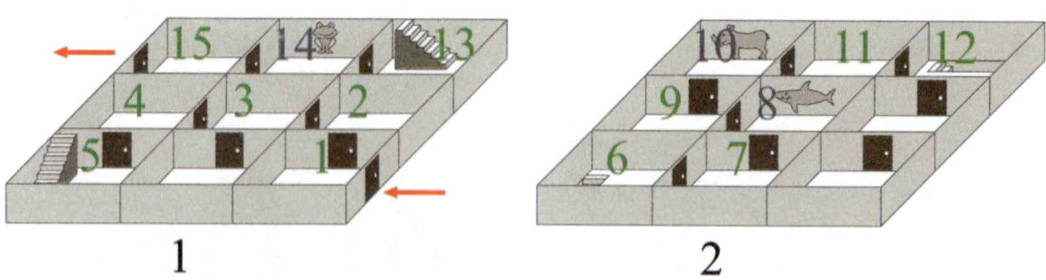

21. (D) 7

Emma finished third, so there are at least 3 dancers in the competition, including her. After Emma we have 3 more dancers and the dancer who is in last place. In total that is be $3 + 3 + 1 = 7$ dancers in the competition.

22. (E) 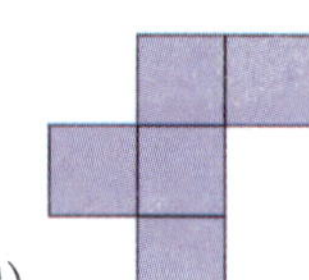

Each piece will cover five numbers. The best case scenario for getting the largest sum is if Malik can cover the five largest numbers: 9, 8, 7, 6, and 5. Piece (E) will allow him to do this.

1	6	7
9	5	4
2	8	3

23. (B) 6

On each of the 9 nights, one frog sang, the other two listened. The first frog sang 2 times; the second frog listened 5 times. She listened to the first frog 2 times, so she much have listened to the third frog 3 times. So, the third frog sang 3 times, and so it must have listened $9 - 3 = 6$ times.

24. (C) 10

With the digits 1, 1, 2, and 3 we can make the expressions: $11-2=9$, $11-3=8$, $12-1=11$, $12-3=9$, $13-1=12$, $13-2=11$, $21-1=20$, $21-3=18$, $23-1=22$, $31-1=30$, $31-2=29$, $32-1=31$. There are 12 expressions but only 10 unique results.

Part III

Answer Keys

	2005	2007	2009	2011	2013
1	A	D	D	C	D
2	C	B	B	D	B
3	E	C	C	D	A
4	D	A	D	A	D
5	C	D	B	B	B
6	C	D	A	B	E
7	C	C	C	C	C
8	D	A	A	B	D
9	C	C	C	C	E
10	C	D	D	A	C
11	A	B	D	D	A
12	A	D	B	B	B
13	B	D	B	B	D
14	C	C	C	A	E
15	A	A	C	B	C
16	A	A	C	C	A
17	D	D	B	A	E
18	B	C	D	D	C
19	B	B	A	D	C
20	C	E	D	D	B
21	E	D	C	B	A
22	A	E	B	A	D
23	D	E	C	D	D
24	E	E	B	B	D

	2015	2017	2019	2021	2023
1	D	D	D	E	D
2	B	C	C	B	B
3	C	B	C	A	A
4	E	A	E	D	A
5	B	A	B	A	E
6	E	A	C	D	C
7	C	E	A	E	B
8	B	C	C	B	D
9	D	D	E	D	D
10	C	A	B	D	C
11	C	D	A	C	C
12	E	E	E	E	E
13	E	D	B	A	C
14	D	A	B	C	B
15	C	C	E	B	D
16	A	B	D	D	A
17	D	B	D	D	A
18	D	C	A	A	C
19	C	C	D	C	A
20	D	E	B	C	B
21	D	E	C	B	D
22	B	D	D	C	E
23	C	B	A	E	B
24	A	E	C	C	C